高等学校土建类学科专业"十三五"系列教材
高等学校系列教材

Hydrology
水　文　学

Li Shanshan　　Li Youyu　　Zhang Shaoliang
　李珊珊　　　李友雨　　　张绍亮　　　主编

中国建筑工业出版社
China Architecture & Building Press

图书在版编目（CIP）数据

水文学＝Hydrology：英文/李珊珊，李友雨，张绍亮主编. —北京：中国建筑工业出版社，2020.10
高等学校土建类学科专业"十三五"系列教材　高等学校系列教材
ISBN 978-7-112-25264-0

Ⅰ. ①水… Ⅱ. ①李… ②李… ③张… Ⅲ. ①水文学-高等学校-教材-英文 Ⅳ. ①P33

中国版本图书馆CIP数据核字（2020）第108238号

  Taking the hydrological cycle as the outline, this book illustrates the physical mechanism, spatial and temporal distribution and the interactions of various elements of the hydrological cycle. It includes basic knowledge of hydrology, theories of hydrological cycle and calculation principles of hydrological balance, analysis of the specific process of hydrological cycle on terrestrial surface water and groundwater, the impact of human activities on the water environment. This book tries to reflect the new ideas and achievements in the field of hydrology.
  It can be used as a textbook for graduates or high-level college students or a reference book for hydrologists, teachers or students in hydrology relevant major.
  本书以水文循环为大纲，阐述了水文循环的物理机制、时空分布及各要素之间的相互作用。包括水文基础知识、水文循环理论和水文平衡计算原理、水文循环对地表水和地下水的具体过程分析、人类活动对水环境的影响。本书反映了水文学领域的新思想和新成果。
  本书既可作为研究生、高层次本科生的教材，也可作为水文学者或水文相关专业师生的参考书。
  本书配备教学课件，有需要的读者可发送邮件至 jiangongkejian@163.com 索取。

责任编辑：赵　莉　吉万旺
责任校对：张　颖

高等学校土建类学科专业"十三五"系列教材
高等学校系列教材
Hydrology
水文学
Li Shanshan　Li Youyu　Zhang Shaoliang
李珊珊　　李友雨　　张绍亮　主编

*

中国建筑工业出版社出版、发行（北京海淀三里河路9号）
各地新华书店、建筑书店经销
霸州市顺浩图文科技发展有限公司制版
北京建筑工业印刷厂印刷

*

开本：787毫米×1092毫米　1/16　印张：11¾　字数：292千字
2020年10月第一版　2020年10月第一次印刷
定价：**46.00**元（赠课件）
ISBN 978-7-112-25264-0
（36049）

版权所有　翻印必究
如有印装质量问题，可寄本社图书出版中心退换
（邮政编码100037）

前　言

水文学是研究地球上水的形成、循环与分布，水的物理、化学性质，以及其与物理和生物环境的相互作用、与人类活动相互关系的科学。水文循环是联系地表大气圈、岩石圈和生物圈的纽带，对地球物理学和自然地理学的发展起着重要的推动作用。此外，水资源是支撑社会经济发展的基本条件。科学认识水文循环规律是合理开发和利用水资源的前提。因此，研究水文学是社会长期发展的现实需求。

本书共有 8 章。第 1 章从总体上论述了水的性质和水资源的概况。第 2 章描述了水文循环的过程和物质守恒理论。接下来的 3 章是以定量的方式描述水文循环的各个要素。第 3 章主要介绍了气象降水的定义、测量和估算方法。第 4 章主要讨论了水在不同介质和界面的蒸发过程与机理。第 5 章主要介绍了径流量的测量和洪水的评估。第 6 章主要包括径流过程线分析和统计学在径流估算中的应用。第 7 章主要描述了地下水理论和地下水流动的规律。第 8 章主要介绍了全球变化中的水资源管理的理念。本书包含了大量的最新降水、径流以及气象方面的其他数据。整篇内容经过严格修订，旨在反映水文学领域的研究和实践现状。

本书中所涉及的水文学理论、原理和过程较容易理解，但需要批判性思维。本书主要面向具有一定的微积分和物理学基础的土木工程、地球科学专业的研究生和优秀本科生。本书可作为水文学相关双语课程的教材和参考书。

致谢

我们感谢聊城大学规划教材项目对本书的资金支持，感谢中国建筑工业出版社赵莉编辑、吉万旺编辑的大力支持，也感谢匿名审稿人对我们的认可。同时，我们还要感谢为我们进行早期校稿的同事们。

Preface

Hydrology is the science, which deals with waters from the Earth, their occurrence, circulation and distribution on the planet, their physical and chemical properties, and their interactions with the physical and biological environment, including their responses to human activities. Hydrological cycle is the link connecting the surface processes of atmosphere, lithosphere and biosphere. It plays an important role in promoting the development of geophysics and physical geography. In addition, water resources are the basic conditions to support social and economic development. Scientific understanding of the law of hydrological cycle is the premise of rational development and utilization of water resources. Therefore, the development of hydrology also has a long-term practical demand for production practice.

There are eight chapters in this book. Chapter 1 discusses the properties of water and water resources in an overall perspective. Chapter 2 describes the process and mass balance of hydrological cycle. The following three chapters describe the elements of the hydrological cycle in a quantitative manner. Chapter 3 covers definition, measurement and estimation of meteorological precipitation approaches. Chapter 4 discusses the evaporation from different media and interfaces. Chapter 5 relates to field practices of streamflow measurements, assessment of extreme flows——floods. Chapter 6 discusses the hydrograph analysis to estimate streamflow and applications of statistical and stochastic processes for estimating streamflows. Chapters 7 explores the theory of groundwater, the application and development of groundwater flow; while Chapter 8 discusses water resources management in a changing world. It contains very extensive data related to precipitation, streamflow and other meteorological aspects. These have been brought up to date. The technical content of the book has been critically reviewed to reflect the current state of practice in the field of hydrology.

The hydrology theories, principles and processes in this book is not hard to understand, but it needs critical thinking most of the time. The coverage presupposes a modest background in calculus and physics. It is aimed at graduate students and upper-level undergraduates majoring in civil engineering or geosciences. It is suitable as a text book or reference for bilingual courses related with hydrology.

Acknowledgements
We are thankful for the opportunity to do this writing project funded by Liaocheng University, an opportunity afforded the freedom of an academic career. We received excellent editorial support from Li Zhao, Wanwang Ji and other staff at China Architecture & Building Press. We are grateful for the care of external reviewers, anonymous to us. We would also like to thank the many colleagues who helped proofread early drafts of the book.

Content

Chapter 1　INTRODUCTION ·· 1
　1.1　The Importance of Water ··· 1
　1.2　Properties of Water ·· 2
　1.3　Water Resources ·· 4
　　1.3.1　Definition ·· 4
　　1.3.2　World's Water Storage ·· 5
　　1.3.3　World's Water Distribution ·· 6
　　1.3.4　Water Resources in China ·· 8
　1.4　Hydrology and Its Branches ·· 10
　Summary ·· 12
　Questions ··· 13

Chapter 2　HYDROLOGICAL CYCLE AND HYDROLOGICAL BALANCE ··············· 14
　2.1　Hydrological Cycle ·· 14
　　2.1.1　The Global Hydrological Cycle ·· 14
　　2.1.2　The Watershed Hydrological Cycle ··· 16
　2.2　Hydrological Budget ·· 18
　Summary ·· 20
　Questions ··· 20

Chapter 3　PRECIPITATION ·· 22
　3.1　Formation of Precipitation ··· 22
　　3.1.1　Atmospheric Cooling ··· 22
　　3.1.2　Condensation Nuclei ··· 23
　　3.1.3　Water or Ice Droplet Growth ··· 23
　　3.1.4　Types of Precipitation ··· 24
　3.2　Distribution of Precipitation ·· 25
　　3.2.1　Static Influences on Precipitation Distribution ·· 25
　　3.2.2　Forest Rainfall Partitioning ··· 26
　3.3　Measurement of Precipitation ·· 28
　　3.3.1　Direct Precipitation Measurement ·· 29
　　3.3.2　Surrogate Precipitation Measurements ·· 33
　3.4　Estimation of Spatially Distributed Precipitation ·· 35
　　3.4.1　Thiessen's Polygons ·· 35
　　3.4.2　Hypsometric Method ··· 36
　　3.4.3　Isohyetal and Other Smoothed Surface Techniques ···································· 37
　3.5　Temporal Characteristics of Precipitation ·· 38
　Summary ·· 41

Translation of Some Sections ··· 41
Questions ·· 51
Chapter 4 EVAPORATION ··· 52
  4.1 Evaporation Mechanisms ·· 52
  4.2 Evaporation from Water Surface ·· 53
  4.3 Evaporation from Soil Surface ·· 55
  4.4 Transpiration ·· 56
  4.5 Evapotranspiration ·· 56
  Summary ·· 57
  Translation of Some Sections ·· 58
  Questions ·· 59
Chapter 5 RUNOFF ··· 60
  5.1 Runoff Mechanism ·· 62
    5.1.1 Overland Flow ·· 62
    5.1.2 Subsurface Flow ·· 63
    5.1.3 Baseflow ·· 65
    5.1.4 Channel Flow ··· 65
  5.2 Velocity Distribution in a Stream Section ·· 66
  5.3 Measuring Streamflow ·· 67
    5.3.1 Instantaneous Streamflow Measurement ··· 67
    5.3.2 Continuous Streamflow Measurement ··· 69
  5.4 Floods ··· 73
    5.4.1 Introduction ·· 73
    5.4.2 Influences on Flood Size ··· 73
  5.5 Watersheds ·· 75
    5.5.1 Watersheds and Types of Drainage Pattern ·· 76
    5.5.2 Watersheds and Stream Orders ··· 78
    5.5.3 A Geomorphologic Perspective ··· 79
    5.5.4 A Topographic Perspective ··· 80
  Summary ·· 81
  Translation of Some Sections ·· 81
  Questions ·· 91
Chapter 6 STREAMFLOW ANALYSIS ·· 92
  6.1 Hydrograph Analysis ·· 92
    6.1.1 The Hydrograph ·· 92
    6.1.2 Hydrograph Separation ·· 93
    6.1.3 The Unit Hydrograph (UHG) ·· 94
  6.2 Flow Duration Curves ·· 97
  6.3 Frequency Analysis ·· 100
    6.3.1 Frequency Distribution ··· 100
    6.3.2 Flood Frequency Analysis ··· 101
  Summary ·· 103

Translation of Some Sections		103
Questions		110

Chapter 7 GROUNDWATER HYDROLOGY ... 112
  7.1 Porous Materials ... 112
  7.2 Aquifer Classification ... 114
    7.2.1 Aquifers and Aquitards ... 114
    7.2.2 Aquifer Characteristics ... 114
    7.2.3 Unconfined or Confined Aquifer ... 115
  7.3 Storage of Groundwater ... 116
    7.3.1 Water in the Unsaturated Zone ... 116
    7.3.2 Water in the Saturated Zone ... 118
  7.4 Surface Tension and Capillarity ... 118
  7.5 Movement of Groundwater ... 119
    7.5.1 Water Transmit in Pores ... 119
    7.5.2 Principles of Groundwater Flow ... 122
    7.5.3 Aquifer Recharge and Discharge ... 124
  Summary ... 128
  Translation of Some Sections ... 128
  Questions ... 138

Chapter 8 WATER RESOURCES MANAGEMENT IN A CHANGING WORLD ... 140
  8.1 Hydrology and Change ... 140
    8.1.1 Climate Change ... 141
    8.1.2 Change in Land Use ... 142
    8.1.3 Groundwater Depletion ... 144
    8.1.4 Urbanization ... 144
  8.2 Water Resource Management ... 147
    8.2.1 Integrated Water Resource Management (IWRM) ... 148
    8.2.2 Integrated Catchment Management (ICM) ... 150
    8.2.3 Stormwater Management ... 151
  Summary ... 154
  Translation of Some Sections ... 154
  Questions ... 159

APPENDIX A: UNITS AND CONVERSIONS ... 160
APPENDIX B: SI UNIT PREFIXES ... 163
APPENDIX C: PHYSICAL PROPERTIES OF WATER (SI UNITS) ... 164
APPENDIX D: GLOSSARY ... 165
REFERENCES ... 179

Chapter 1 INTRODUCTION

1.1 The Importance of Water

Water is the most common substance on the surface of this planet. It is one of the few substances that can be found in all three states (gas, liquid and solid) within the earth's climatic range. The role of water is crucial to most natural processes.

The very presence of water in all three forms makes it possible for the earth to have a climate that is habitable for life forms: water acts as a climate ameliorator through the energy absorbed and released during transformation between the different phases. In addition to lessening climatic extremes, the transformation of water between gas, liquid and solid phases is vital for the transfer of energy around the globe——moving energy from the equatorial regions towards the poles. The low viscosity of water makes it an extremely efficient transport agent, whether through international shipping or river and canal navigation. These characteristics can be described as the physical properties of water and they are critical for human survival on planet earth. The chemical properties of water are equally important for our everyday existence. Water is one of the best solvents naturally occurring on the earth. This makes water vital for cleanliness: we use it for washing but also for the disposal of pollutants. The solvent properties of water allow the uptake of vital nutrients from the soil and into plants; this then allows the transfer of the nutrients within a plant's structure. The ability of water to dissolve gases such like oxygen allows life to be sustained within bodies of water such as rivers, lakes and oceans. The capability of water to support life goes beyond bodies of water; the human body is composed of around 60% water. The majority of this water is within cells, but there is a significant proportion (around 34%) that moves around the body carrying dissolved chemicals, which are vital for sustaining our lives. Our bodies can store up energy reserves that allow us to survive without food for weeks but not more than days without water. There are many other ways that water affects our very being. In places such as Norway, parts of the USA and New Zealand, energy generation for domestic and industrial consumption is through hydroelectric schemes, harnessing the combination of water and gravity in a largely sustainable manner. The study of water would therefore also seem to underpin our very existence.

Before expanding further study of hydrology, it is first necessary to step back and take a closer look at the properties of water briefly outlined above. Even though water is the most common substance found on the earth's surface, it is one of the strangest.

1.2 Properties of Water

Water is a unique molecule, present in three physical states and in bulk quantities on the earth. In the liquid state, these water molecules and ions are closely packed, but constantly moving and jostling each other. The physical properties of water have their basis in these molecular-scale interactions, so a brief mention of water's molecular properties is relevant here. Water molecules are polar, with more positive charge near the hydrogen atoms and more negative charge near the oxygen atom, as shown in Figure 1-1. This uneven charge distribution causes attraction among the hydrogen atoms of one molecule and the oxygen atoms of another (Figure 1-2). The polarity and self-attraction of water molecules is the fundamental cause of its remarkable properties:

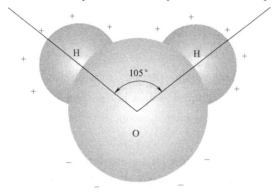

Figure 1-1　Geometry of a Water Molecule
Source: FITTS C R. Groundwater Science, 2002.

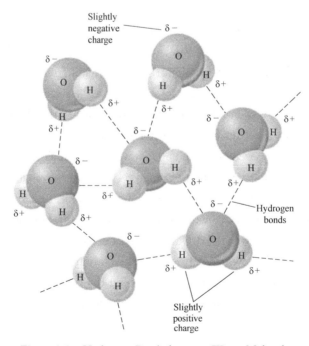

Figure 1-2　Hydrogen Bonds between Water Molecules
Source: MILLER G T, SPOOLMAN S E. Environmental Science (14th ed). 2013.

(1) Forces of attraction, called hydrogen bonds, hold water molecules together, which is the major factor determining water's distinctive properties.

(2) Water exists as a liquid over a wide temperature range because of the forces of attraction between its molecules. Without water's high boiling point, the oceans would have evaporated long ago.

(3) Liquid water changes temperature slowly because it can store a large amount of heat without a large change in its own temperature (Table 1-1). This high heat storage capacity helps protect living organisms from temperature changes, moderates the earth's climate, and makes water an excellent coolant for car engines and power plants.

Specific Heat Capacity of Various Substances　　　　　　　　　　Table 1-1

Substance	Specific Heat Capacity[kJ/(kg · K)]
Water	4.2
Dry soil	1.1
Ethanol (alcohol)	0.7
Iron	0.44

(4) It takes a large amount of energy to evaporate water because of the forces of attraction between its molecules. Water absorbs large amounts of heat as it changes into water vapor and releases this heat as the vapor condenses back to liquid water. This helps to distribute heat throughout the world and to determine regional and local climates. It also makes evaporation a cooling process, which explains why you feel cooler when perspiration evaporates from your skin.

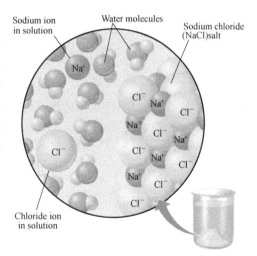

Figure 1-3　How Salt Dissolves in Water
Source: MILLER G T, SPOOLMAN S E. Environmental Science (14th ed). 2013.

(5) Liquid water can dissolve a variety of compounds (Figure 1-3). It carries dissolved nutrients into the tissues of living organisms, flushes waste products out of those tissues, serves as an all-purpose cleanser, and helps to remove and dilute the water-soluble wastes of civilization. This property also means that water-soluble wastes can easily pollute water.

(6) Water filters out wavelengths of the sun's ultraviolet radiation that would harm some aquatic organisms. However, down to a certain depth, it is transparent to sunlight that

is necessary for photosynthesis. The forces of attraction between water molecules also allow liquid water to adhere to a solid surface. This enables narrow columns of water to rise through a plant from its roots to its leaves (a process called capillary action).

(7) Unlike most liquids, water expands when it freezes, as shown in Figure 1-4. This means that ice floats on water because it has a lower density (mass per unit of volume) than liquid water has. Otherwise, lakes and streams in cold climates would freeze solid, losing most of their aquatic life. Because water expands upon freezing, it can break pipes, crack a car's engine block (if it does not contain antifreeze), break up pavement, and fracture rocks. In water's solid state (i. e. ice), the hydrogen bonds become rigid and a three-dimensional crystalline structure forms. The maximum density of water actually occurs at around 4℃ (Figure 1-5) so that still bodies of water such as lakes and ponds will display thermal stratification, with water close to 4℃ sinking to the bottom.

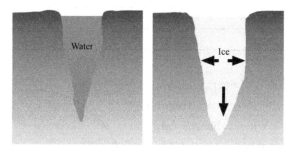

Figure 1-4 The Volume Expands when Water Freezes

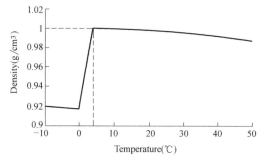

Figure 1-5 The Density of Water with Temperature

Note: The broken line shows the maximum density of water at 3.98℃.

1.3 Water Resources

1.3.1 Definition

The reason why water belongs to resource is determined by its physical, chemical and natural characteristics. There are two definitions of water resources, generalized and specialized.

(1) Generalized Water Resources
Water resources in a generalized concept refer to the water widely existing on the earth's surface and in the earth's lithosphere, atmosphere and biosphere in the form of solid, liquid and gas in nature, which is the total water quantity of the earth including seawater. The earth we live on is a planet of water, oceans, rivers, lakes and streams. Through interaction between the elements of hydrological cycles, a complete earth hydrosphere is formed, which is also the origin of life. In this sense, any water in the hydrosphere has direct or indirect use value for human beings and could be regarded as water resources. It includes renewable water resources and non-renewable water resources.

(2) Specialized Water Resources
Water resources in a specialized concept are the water enriched in rivers, lakes, glaciers and buried in shallow aquifers, which can be replenished and renewed by the cycling of water. It is easy to be exploited by human beings, including surface water, groundwater and soil water. Among them, the surface water is rivers, glaciers, lakes, swamps and other water bodies; the groundwater is the dynamic water volume of the underground catchment; the soil water is the water dispersed in the loose surface layer of the lithosphere.

1.3.2 World's Water Storage
The oceans covers up to 70% of this planet. Groundwater has a volume of $8.4 \times 10^6 km^3$; nevertheless, half of it cannot be exploited because either it is located so deep that pumping is not economically feasible or it is saline. From the exploitable amount of the planet's water, river and lake water comprise only a small fraction of almost 2%. The remaining 98% of the exploitable quantity of water is groundwater. From these arguments, it is clear that a minimal quantity of groundwater is crucial for satisfying human needs. The recharge of a significant part of groundwater is done via the infiltration of the atmospheric precipitation, while on the other hand, substantial amount of runoff ends into the oceans and seas.

Water exists in virtually every accessible environment on or near the earth's surface. It is in animal blood, trees, air, glaciers, streams, lakes, oceans, rocks and soil. Its distribution among the main reservoirs is listed in Table 1-2. Of the fresh water reservoirs, glacial ice and groundwater are by far the largest. Groundwater and surface water are the two reservoirs most used by humans because of their accessibility. Fresh groundwater is about 100 times more plentiful than fresh surface water, but we use more surface water because it is so easy to find and use. Much of the total groundwater volume is deep in the crust and too saline for most uses, as shown in Figure 1-6.

Storage of Water in Earth's Reservoirs Table 1-2

Reservoir		Percent of All Water(%)	Percent of Fresh Water(%)
Oceans		96.5	
Ice and snow		1.8	69.6
Groundwater	Fresh	0.76	30.1
	Saline	0.93	
Surface water	Fresh lakes	0.007	0.006
	Saline lakes	0.26	
Marshes		0.0008	0.03
Rivers		0.0002	0.006
Soil moisture		0.0012	0.05
Atmosphere		0.001	0.04
Biosphere		0.0001	0.003

Source: MAIDMENT D R. Handbook of Hydrology. 1993.

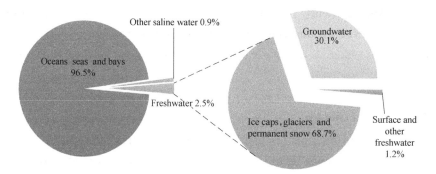

Figure 1-6　The Storage of Freshwater on Earth

Fueled by energy from solar radiation, water changes phase and cycles continuously among these reservoirs in the hydrologic cycle (Figure 2-1). Solar energy drives evaporation, transpiration, atmospheric circulation and precipitation. Gravity pulls precipitation down to earth, pulls surface water and groundwater down to lower elevations, and ultimately back to the ocean reservoir.

1.3.3　World's Water Distribution

Global freshwater resources are not only in short supply, but also unevenly distributed. According to its regional distribution, 9 countries (Brazil, Russia, United States, Canada, China, Colombia, Indonesia, Peru and India) possess 60% of the world's freshwater resources; while 80 countries and regions, which have 40% of the world's total population, are in a shortage of freshwater (Figure 1-7). 300 million people in 26 countries live in absolute water scarcity. It is estimated that, by 2025, 3 billion people will suffer from water shortage, which involves more than 40 countries and regions.

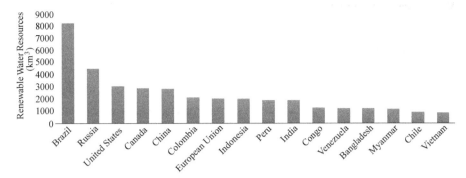

Figure 1-7　Total Renewable Water Resources Ranking in 2011

Water resources are becoming a kind of scarce resources in the 21st century. The problem of water resources is not only a resource problem, but also a major strategic problem related to the sustainable development of national economy, society and long term stability. Since the beginning of the 20th century, with the population expansion and the rapid expansion of industrial and agricultural production scale, the global freshwater consumption has grown rapidly. From 1900 to 1975, the world's agricultural water consumption has increased by 7 times, and the industrial water consumption has increased by 20 times. In recent decades, the water consumption is continuously increasing at the rate of 4%~8% per year, and the contradiction between the supply and demand of fresh water has become increasingly prominent.

Water scarcity can mean scarcity in availability due to physical shortage, or scarcity in access due to the failure of institutions to ensure a regular supply or due to a lack of adequate infrastructure. Water stress starts when the water available in a country drops below 1700m^3 per capita. When the 1000m^3 per capita threshold is crossed, water scarcity is experienced. Absolute water scarcity is considered for countries with less 500m^3 per capita. By this definition, 49 countries are water stressed, 9 of which experience water scarcity and 21 are in absolute water scarcity.

To strengthen the research on the development, utilization and management, water resources protection has been put on the agenda and developed rapidly. The United Nations (UN), Food and Agriculture Organization (FAO), World Meteorological Organization (WMO), United Nations Educational, Scientific and Cultural Organization (UNESCO), and United Nations Industrial Development Organization (UNIDO) all have research projects on water resources and continuously organize international exchanges. In 1965, UNESCO established the International Hydrological Decade (IHD), with more than 120 countries participating in water resources research. It is a global-scale challenge in the

21st century.

1.3.4 Water Resources in China

1.3.4.1 Storage of Water Resources
China's water resource storage is $2.8 \times 10^{12} m^3$. Among them, the surface water is $2.7 \times 10^{12} m^3$, and the underground water is $0.83 \times 10^{12} m^3$. Due to the recharge and disrecharge of surface water and underground water, $0.73 \times 10^{12} m^3$ is double counted, therefore, the amount of underground water resource should be adjusted to $0.1 \times 10^{12} m^3$.

The total amount of water resources in China is abundant, but the per capita share is low. According to the world bank's continuous statistics, China's total water resources rank 6th in the world, with a per capita share of $2240 m^3$, which is about 1/4 of the world's per capita, ranking 88th out of 153 countries.

At present, the per capita water resources (excluding the transit water) of 16 provinces (regions and cities) in China are water stressed, and the per capita water resources of 6 provinces (regions), including Ningxia, Hebei, Shandong, Henan, Shanxi and Jiangsu, are lower than $500 m^3$, which are in absolute water scarcity.

1.3.4.2 Regional Distribution of Water Resources
Due to geographical location, water vapor source, topographic conditions and other factors, the regional distribution of water resources in China is very uneven, with the general trend of decreasing from the southeast coast to the northwest inland. This is not matched with the distribution of soil resources in China. The drainage area of Yangtze River and its south accounts for 36.5% of the whole country, while its water resources account for 81% of the total amount in China; the drainage area of the Huaihe River and its north accounts for 63.5% of the whole country, while its water resources account for 19% of the total amount in China.

Precipitation is an important source of water supply. According to the amount of annual precipitation and annual runoff, our country can be roughly divided into five zones with different water resources conditions:

(1) Water-rich Zone
The annual precipitation is more than 1600mm; the annual runoff depth is more than 800mm; and the annual runoff coefficient is more than 0.5. It includes most areas of Zhejiang, Fujian, Taiwan and Guangdong, eastern areas of Guangxi, southwestern areas of Yunnan, southeastern areas of Tibet, and the mountainous areas of Jiangxi, Hunan and western areas of Sichuan.

Among them, the northeastern part of Taiwan and the southeastern part of Tibet have the most abundant water resources in China, with the annual runoff as high as 5000mm.

(2) Humid Zone
The annual precipitation is 800~1600mm; the annual runoff depth is 200~800mm; and the annual runoff coefficient is 0.25~0.5. It mainly includes the lower drainage area of Yishu River and Huaihe River, the south part of Hanshui drainage area, the east part of Yangtze River drainage area, most part of Yunnan, Guizhou, Sichuan and Guangxi provinces, and the northeast part of Changbai Mountain area.

(3) Sub-humid Zone
The annual precipitation is 400~800mm; the annual runoff depth is 50~200mm; and the annual runoff coefficient of 0.10~0.25. It includes the Huang-Huai-Hai Plain, Heilongjiang, Jilin, Liaoning, Shanxi, most of Shaanxi, the southeastern part of Gansu and Qinghai provinces, the northern and western mountain areas of Xinjiang, the northwestern part of Sichuan and the eastern part of Tibet.

(4) Semi-arid Zone
The annual precipitation is 200~400mm; the annual runoff depth is 10~50mm; and annual runoff coefficient is below 0.1. It includes most parts of Inner Mongolia, Ningxia, Qinghai and Gansu, the northwestern part of Xinjiang, and some parts of Tibet.

(5) Arid Zone
The annual precipitation is less than 200mm, and the annual runoff depth is less than 10mm. Some areas are flowless areas, including deserts in Inner Mongolia, Ningxia and Gansu, Qaidam Basin in Qinghai, Tarim Basin and Junggar Basin in Xinjiang, and Qiangtang region in Tibet.

Due to the differences in precipitation and hydrogeological conditions, groundwater resources in the plain area vary greatly. The regional distribution of runoff is similar as that of precipitation, but the distribution is more uneven due to the influence of land surface.

High plateaus in the west and low plains in the east characterize the topography of China. Under such topographic conditions, the Chinese rivers mostly rise in high mountains and plateaus, running rapidly downward and flowing into the sea through steeper channels. Many rivers in China have a total hydraulic drop of over 1000m. Of the major rivers (Yangtze River, Yellow River, Yarlung Zangbo River, Lancang River and Nujiang River), total hydraulic drop is 2000~5000m.

1.3.4.3 Temporal Distribution of Water Resources

The interannual or annual distribution of water resources in China is uneven, which is heavily influenced by regional climate.

The variations of interannual precipitation is great, years of droughts and floods occur continuously. Many rivers have periods of high or low flow lasting for 3~8 years. For example, the Yellow River was in low flow for 11 consecutive years from 1922 to 1932, and in high flow for 9 consecutive years from 1943 to 1951. The discrepancy of the maximum and the minimum annual precipitation is huge in China. The maximum annual precipitation in the south is 2~4 times that of the minimum, and the maximum annual precipitation in the north is 3~6 times that of the minimum. For example, the precipitation in Beijing was 1405mm in 1959, 5.5 times of which in 1921 (256mm).

The annual precipitation is also not evenly distributed. The precipitation is most frequent in summer and least frequent in winter. Spring and autumn ranks between them. Spring rain is mainly caused by cyclones passing through, while autumn rain is mainly caused by typhoons passing through. Take Beijing as an example, the precipitation from June to September accounted for 80% of the annual precipitation.

It is precisely because of the uneven distribution of precipitation, some places have more water resources in a certain time, while others have less water resources. Thus, it is urgent to apply joint management of surface water and groundwater in China.

1.4 Hydrology and Its Branches

According to UNESCO, hydrology is the science which deals with waters from the earth, their occurrence, circulation, distribution on the planet, their physical and chemical properties, and their interactions with the physical and biological environment, including their responses to human activities. Hydrology is one of the basic subjects, which is also closely related to climatology and micrometeorology, soil science, geology and geophysics, ecology, and geography geology.

Hydrology could be branched as marine hydrology, hydrometeorology, terrestrial hydrology and applied hydrology. Marine hydrology focuses on the study of chemical composition and physical properties of seawater, waves, tides and currents in the ocean, and lateral sediment movement along the coast. Hydrometeorology studies the relationship between hydrosphere and atomsphere, including the hydrologic cycle and hydrologic balance in the atmosphere, which lays the foundation for the theory of water exchange through transpiration, condensation and precipitation, and the occurrence and formation of storms or droughts. Terrestrial hydrology studies all aspects of terrestrial water bodies. Applied hydrology, which serves production directly, is the most vigorous branch of modern hydrology. Hydrology usually refers primarily to terrestrial hydrology and applied hydrology.

The development of hydrology experienced four periods:

(1) Embryonic Period (from Ancient Times to the 1500s)
In this period, people began some primitive hydrologic observation and started to accumulate the original hydrological knowledge.

In 2300BC, Chinese began to observe the fluctuation of river stage. The "Stone man" of Dujiangyan set up by Li Bing of Qin Dynasty, together with "Water rule" inscriptions set up in Sui Dynasty, and the "water rule tablet" set up in Song Dynasty, illustrate the continuous progress of river stage observation. Rainfall observations first appeared in India in 400BC. In Qin Dynasty (221BC~207BC), China developed a system of measuring and reporting rainfall. In 1247, a more advanced method for calculating rainfall depth had been developed, and the "bamboo device for measuring snow" was used to obtain snowfall depth. In the 5th century, China began to use "floating bamboo" to measure river velocity. In Ming Dynasty, a sand sampling device was invented to determine the amount of sediment in the Yellow River.

In 239BC, the book *Spring and Autumn Annals of Lv* had recorded primary idea of hydrological cycle. At the beginning of the 6th century, *Commentary on the Waterways Classic* recorded the general state of 1252 rivers in the territory of China at that time, which was one of the earliest hydrogeological surveys.

(2) Foundation Period (the 1500s~the 1900s)
Modern hydrological instruments have brought the hydrological observation into a scientific quantitative stage. In 1663, Ryan and Hooke invented the self-recording tipped-bucket rain gauge; in 1687, Halley invented the evaporator to measure the evapotranspiration of water surface; in 1870, Ellis invented the rotary cup current meter; and in 1885, Price invented pull-type velocimetry.

Later, various hydrographic stations began to appear. The theory of modern hydrology developed gradually. In 1674, Perot made a comparative analysis on runoff and rainfall of the Seine River, and put forward the concept of water balance, which later developed into one of the basic principles of hydrology. In 1738, Bernoulli published his hydroenergy equation. In 1775, Dzeicai published his uniform flow formula in open channel. In 1802, Dalton developed the Dalton Equation for calculating evaporation from water surface. In

1851, Molvaney proposed the concepts of confluence and confluence coefficient, and the famous reasoning formula for calculating the maximum flow. In 1856, Darcy published Darcy's Law describing the movement of groundwater in homogeneous media. Subsequently, monograph on hydrology began to appear, which laid a foundation for hydrology as a modern science.

(3) The Formation of Applied Hydrology (the 1900s~the 1950s)

Stepping into the 20th century, the demand of flood control, irrigation, hydropower generation and forestry has put forward more and more new issues to hydrology, and the methods to solve these issues were gradually theorized and systematized. During 1914 ~ 1924, probability theory, mathematical statistics theory and methods were introduced into hydrology systematically.

In 1932~1938, Sherman, Horton, McCarthy, Snyder, Clark and Linsley pioneered on the method of unit hydrograph, combining analysis of multiple hydrological variables and runoff regulation calculation. During this period, hydrology stations developed into networks of national-level hydrology stations. These achievements have laid the foundation and taken the lead on engineering hydrology. Other branches of applied hydrology, such as agricultural hydrology, forest hydrology and urban hydrology emerged subsequently. In 1949, books of *Applied Hydrology*, *Principles of Applied Hydrology*, *Hydrology Manual* and other applied hydrology monographs were published, which marked the formation of applied hydrology.

(4) Modern Hydrology Period (since the 1950s)

Since the 1950s, the rapid development of science and technology thrived the relationship between human beings, and the studies on water have entered into a new era. It gives the hydrology new characteristics: 1) Water resources are highly demanded. Hydrology is the guidance to explore surface water and groundwater for the development and utilization of water resources. 2) Human activities on the hydrological cycle impact the earth's environment. Environmental hydrology, an interdisciplinary subject of hydrology and environmental science, has been conceived and formed. 3) Modern science and technology, such as nuclear technology, remote sensing, computing technology, theory and method of systematic analysis, are penetrating into every field of hydrology, which makes a spurt of progress on hydrology. 4) The research field of hydrology is expanding, and the gaps between hydrology and other earth science disciplines have been gradually filled. Interdisciplinary hydrology will thrive in the future.

Summary

Hydrology may be broadly defined as "water science". Therefore, one's background, interests or motives are probably in some way responsible for one's perception of hydrology. In truth, hydrology can be about a rather large variety of subjects. Those whose background is in earth science will probably already be familiar with the important role of groundwater in a variety of geological processes, or how river flow influences and is influenced by the topography. Others may be interested in issues of water resources, protection or water rights. Because water is essential to life, biologists and ecologists also need to understand temporal and spatial distributions of water near the land surface.

Questions

1-1 Discuss the nature of water's physical properties and how important these are in determining the natural climate of the earth.

1-2 How may water-poor countries overcome the lack of water resources within their borders?

1-3 What is the current situation of water resources in China and how may China deal with the water crisis?

Chapter 2 HYDROLOGICAL CYCLE AND HYDROLOGICAL BALANCE

2.1 Hydrological Cycle

The movement of water between the land, oceans and the atmosphere is called the hydrologiccal cycle.

Solar energy drives the hydrological cycle; gravity and other forces also play important roles. The dynamic processes of water vapor formation and transport of vapor and liquid in the atmosphere are driven largely by solar energy. Precipitation and the flow of water on and beneath the Earth's surface are driven primarily by gravity. Within partially dry soil, gravitational and other forces are responsible for the movement of water.

The location at which water may be stored include the ocean and seas, glaciers and other ice, the atmosphere, lakes, reservoirs, streams and rivers, wetlands, soils, geologic formations, plants and animals, and man-made structures. Residence times may range from seconds to thousands of years. Hydrologic processes by which water moves through the hydrological cycle includes atmosphere movement of air masses, precipitation, evaporation, transpiration, infiltration, percolation, groundwater flow, surface runoff and streamflow.

Water quality also changes within the cycle. For example, seawater is converted to freshwater through evaporation. Salts are dissolved from the land by surface and subsurface water. A myriad of other dissolved and suspended constituents enter and leave the water during various phases of the cycle.

The hydrological cycle is a conceptual model of how water moves around between the earth and atmosphere in different states as a gas, liquid or solid. As with any conceptual model, it contains many gross simplifications; these are discussed in this section. There are different scales that the hydrological cycle can be viewed at, but it is helpful to start at the large global scale and then move to the smaller hydrological unit of a river basin or watershed.

2.1.1 The Global Hydrological Cycle

Energy from the sun results in evaporation of water from ocean and land surfaces and causes different heating and resultant movement of air masses. Water vapor is transported with the air masses and under the right conditions it becomes precipitation. Evaporation from the ocean is the primary source of atmospheric vapor for precipitation, but evaporation from soil, streams and lakes and transpiration from vegetation also contribute. Precipitation runoff from the land becomes streamflow. Soil moisture replenishment, groundwater storage and subsurface flow occur because of water infiltrating into the ground. Stream and groundwater flow convey water back to the oceans.

The global process of water exchange provides some stability in the distribution of waters between the land, the oceans and the atmosphere (Figure 2-1). This equilibrium is relative and can change in time, and these changes can lead to corresponding changes in hydrological and climatic conditions.

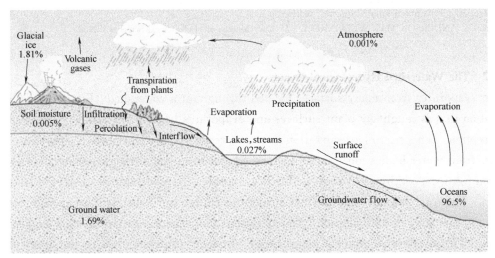

Figure 2-1 Global Hydrological Cycle (The percentages correspond to the volume in each of the different compartments)
Source: MAIDMENT D R. Handbook of Hydrology. 1993.

Water is transferred between reservoirs primarily via four fluxes: precipitation, evapotranspiration, sublimation and runoff. Over land, the runoff ratio represents the fraction of precipitation that contributes to runoff (as shown in Table 2-1).

Estimates of Average Global Annual Precipitation, Evaporation, Runoff Rate, and Runoff Ratio

Table 2-1

Region	Surface Area ($10^6 km^2$)	Precipitation(P) (mm/a)	Evaporation(E) (mm/a)	P−E[①] (mm/a)	(P−E)/P[②]
Europe	10	657	375	282	0.43
Asia	44.1	696	420	276	0.4
Africa	29.8	696	582	114	0.16
Australia	8.9	803	534	269	0.33
North America	24.1	645	403	242	0.38

Continued

Region	Surface Area ($10^6 km^2$)	Precipitation(P) (mm/a)	Evaporation(E) (mm/a)	P−E[①] (mm/a)	(P−E)/P[②]
South America	17.9	1564	946	618	0.4
Antarctica	14.1	169	28	141	0.83
All Land Areas	148.9	746	480	266	0.36
Arctic Ocean	8.5	97	53	44	0.45
Atlantic Ocean	98	761	1133	−372	−0.49
Indian Ocean	77.7	1043	1294	−251	−0.24
Pacific Ocean	176.9	1292	1202	90	0.07
All Oceans	361.1	1066	1176	−110	−0.10
Globe	510	973	973	0	0

Note: ① means runoff rate; ② means runoff ratio.
Source: ANDERSON M G, MCDONNELL J J. Encyclopedia of Hydrological Sciences. 2006.

2.1.2 The Watershed Hydrological Cycle

Water evaporates from many surfaces located throughout a watershed (Figure 2-2). Precipitation that is caught by plant surfaces and evaporates back to the atmosphere is called interception. This part of precipitation does not reach the soil surface. Evaporation also occurs from water bodies located in a watershed and from soil surfaces. Water that is extracted from the soil by plant roots and that evaporates from within plant leaves is transpiration. The total amount of water that evaporates from a watershed is evapotranspiration, which is the sum of interception, transpiration and evaporation from soils and water bodies. This evaporated water is temporarily lost from the watershed to the atmosphere but eventually returns to the earth's surface as precipitation at some other locations and the cycle continues. Precipitation falling on a watershed that is not returned to the atmosphere via evapotranspiration can either flow over the soil surface, reaching stream channels as overland flow or surface runoff, or it infiltrates into the soil.

The fate of infiltrated water depends on the moisture status of the soil, the water holding capacity of the soil, and the network and size of pores within the soil matrix. Infiltrated water that is in excess of the soil water holding capacity can flow downward under the influence of gravity until reaching groundwater. If the downward drainage of water, called percolation, reaches strata of soil or rock with limited permeability, water can be diverted laterally through the soil and discharge into a stream channel or other surface water body as subsurface flow, which is often called interflow. The fate of water that reaches groundwater depends on subterranean characteristics of earth materials and geologic strata that influence the pathways by which groundwater can flow. Some groundwater intersects river channels or other water bodies, thereby returning to surface waters. Water that seeps into deep groundwater aquifers can be stored for centuries before returning to surface waters. Therefore, the journey of water falling as precipitation on watersheds can follow a myriad of pathways to the ocean only to be evaporated and recycled again.

Looking at the relative magnitude of these water fluxes (Figure 2-3), we observe that over the continents evapotranspiration is on average smaller than precipitation. The fraction of precipitation exceeding evapotranspiration contributes to surface and groundwater flow.

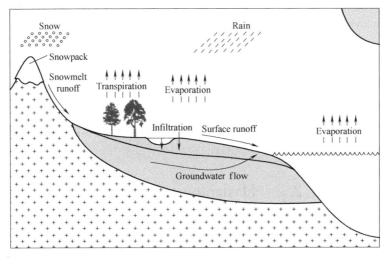

Figure 2-2 Watershed Hydrologic Cycle
Source: FITTS C R. Groundwater Science. 2002.

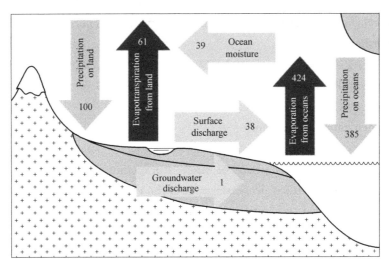

Figure 2-3 Flows within the Hydrological Cycle
Note: Units are relative to the annual precipitation on the land surface ($100=119000km^3$ per year).
Black arrows depict flows to the atmosphere, gray arrows depict flows to land or oceans, and blue (horizontally) arrows indicate lateral flows.
Source: FITTS C R. Groundwater Science. 2002.

Conversely, over the oceans evaporation exceeds precipitation. Therefore, precipitation alone is not sufficient to remove all the water vapor coming from ocean evaporation from the atmosphere above the oceans. In fact, the atmospheric circulation transports part of this water vapor on land masses, thereby allowing continental precipitation to exceed the evapotranspiration rate. It is instructive to analyze the water cycle at different scales. We have seen that at the global scale water reaches earth's surface as precipitation.

A fundamental concept of the hydrological cycle is that water is neither lost nor gained from the Earth over time. However, the quantities of water in the atmosphere, soils, groundwater, surface water, glaciers and other components are constantly changing because of the dynamic nature of the hydrological cycle. A few facts to remember about the hydrological cycle are as follows: Solar energy provides the energy that drives and sustains the cycling of water on Earth. There is no beginning or end to the cycle. Biosphere partitions this water into runoff, soil and groundwater storage, groundwater seepage, or evapotranspiration back to the atmosphere.

2.2 Hydrological Budget

The hydrological cycle is complex and dynamic but can be simplified if we categorize components into input, output and storage (Figure 2-3). Take any region in space, over any period of time, all inputs of water to a watershed must balance with outputs from the watershed and changes in storage of water within the watershed. This hydrologic balance, or water budget, follows the principle of the conservation of mass law expressed by the equation of continuity:

$$I - O = \Delta S \tag{2-1}$$

where I——the inflow;

O——the outflow;

ΔS——the change in storage.

If inflows exceed outflows over a period of time, storage must increase, or if outflows exceed inflows then, storage must decrease——it is the conservation of mass law. The water budget can be expanded to identify the key inputs and outputs for a watershed over a specific period of time as follows:

$$P + \Delta G + \Delta Q - E = \Delta S \tag{2-2}$$

where P——the volume generated by precipitation;

ΔG——the net groundwater discharge volume; $\Delta G = G_i - G_o$: G_i is the groundwater flow into the watershed and G_o is the groundwater flow out of the watershed;

ΔQ——the net volume of streamflow from the watershed; $\Delta Q = Q_i - Q_o$: Q_i is the streamflow flowing into the watershed and Q_o is the streamflow flowing out of the watershed;

E——the evapotranspiration;

ΔS——the change in the amount of storage in the watershed; $\Delta S = S_2 - S_1$; S_2 is the storage at the end of a period and S_1 is the storage at the beginning of a period.

Note that in the water budget of a watershed (Equation 2-2), groundwater flows only need to be taken into account if they contribute to, or diminish surface water on the watershed. Therefore, deep groundwater exchanges that have no contact with surface water are not considered. In some situations, there can be surface water seepage to deep groundwater aquifers, which is a net loss of water from the watershed, expressed as G_o.

This is a volume balance, but because water is so incompressible, it is essentially a mass balance as well. Hydrologic balance is useful for estimating unknown fluxes in many different hydrologic systems. Because the mass of the element in hydrological cycle is constantly on the change, the water budget can also be rewritten as:

$$A\frac{dP}{dt} + \frac{dG}{dt} + \frac{dQ}{dt} - \frac{dE}{dt} = \frac{dS}{dt} \qquad (2\text{-}3)$$

where dP/dt——the precipitation [L/T];

dS/dt——the change in reservoir volume per time [L^3/T];

dG/dt——the net groundwater discharge [L^3/T];

dQ/dt——the net volume of streamflow from the watershed [L^3/T];

dE/dt——the evapotranspiration rate [L/T].

[Example 1-1] Consider a reservoir with one inlet stream, one outlet at a dam and a surface area of 2.5km^2. There hasn't been any rain for weeks, and the reservoir level is falling at a rate of 3.0mm/d. The average evaporation rate from the reservoir surface is 1.2mm/d, the inlet discharge is 10000m^3/d, and the outlet discharge is 16000m^3/d. Assuming that the only other important fluxes are the groundwater discharges in and out of the reservoir. What is the total net rate of groundwater discharge into the reservoir?

[Solution] In this case, the reservoir is the region for which a balance is constructed. Fluxes into this region include the inlet stream flow (Q), and the net groundwater discharge (G). Fluxes out of this region include the outlet stream flow (O) and evaporation (E) from the surface. According to Equation (2-3), the hydrologic balance in this case requires

$$\frac{dG}{dt} + \frac{dQ}{dt} - \frac{dE}{dt} = \frac{dS}{dt}$$

Calculate the rate of change in reservoir volume as:
$$\frac{dS}{dt} = -0.003 \times 2.5 \times 1000^2 = -7500 \text{m}^3/\text{d}$$

Similarly, the rate of evaporative loss is
$$\frac{dE}{dt} = 0.0012 \times 2.5 \times 1000^2 = 3000 \text{m}^3/\text{d}$$

Therefore,
$$\frac{dG}{dt} = \frac{dS}{dt} - \frac{dQ}{dt} + \frac{dE}{dt}$$
$$= -7500 - (10000 - 16000) + 3000$$
$$= 1500 \text{m}^3/\text{d}$$

Solving the first equation for groundwater yields a net groundwater discharge of $1500 \text{m}^3/\text{d}$.

Summary

The hydrological cycle describes the movement and conservation of water on earth. This cycle includes all of the water present on and in the Earth, including salt and fresh water, surface and groundwater, water present in the clouds and that trapped in rocks far below the Earth's surface. This chapter is organized around the unifying concepts of the hydrological cycle and the watershed as a basic unit of study. The goal of physical hydrology is to explain phenomena of water flow in the natural environment by application of physical principles. Solutions to many hydrological problems require an understanding of the dynamics of water motion. Thus, much of the remainder of this book is devoted to the quantitative description of components of the hydrological cycle based on models that arise from fluid dynamics.

Questions

2-1 What is the driven force of hydrological cycle?

2-2 Describe how the hydrological cycle varies around the globe.

2-3 One lake has a surface area of 708000m^2. Based on collected data, one brook flows into the lake at an average rate of $1.5 \text{m}^3/\text{s}$ and another river flows out of this lake at an average rate of $1.25 \text{m}^3/\text{s}$ during the month of June. The evaporation rate was measured as 19.4cm per month. Evapotranspiration can be ignored because there are few water plants on the shore of the lake. A total of 9.1cm of precipitation fell this month. Seepage is negligible. Due to the dense forest and the gentle slope of the land surrounding the lake, runoff is also negligible. The average depth in the lake on June 1st was 19m. What was the average depth on June 30th?

2-4 One watershed, with an area of 4530km^2, receives 77.7cm of precipitation per year. The average rate of flow measured in the river, which drained the watershed, is $39.6 \text{m}^3/\text{s}$. Infiltration was estimated to occur at an average rate of $9.2 \times 10^{-7} \text{m}^3/\text{s}$.

Evapotranspiration is estimated to be 45cm/a. What is the change in storage in the watershed?

2-5 One lake is approximately 12km in length by 2.5km in width. The inflow for the month of April is 3.26m^3/s and the outflow is 2.93m^3/s. The total monthly precipitation is 15.2cm and the evaporation is 10.2cm. The seepage is estimated to be 2.5cm. Estimate the change in storage during the month of April.

2-6 Consider a reservoir with one inlet stream, one outlet at a dam and a surface area of 3.5km^2. There has not been any rain for weeks, and the reservoir level is falling at a rate of 3.0mm/d. The average evaporation rate from the reservoir surface is 1.8mm/d, the inlet discharge is 8000m^3/d, and the outlet discharge is 18000m^3/d. Assuming that the only other important fluxes are the groundwater discharges in and out of the reservoir, What is the total net rate of groundwater discharge into the reservoir?

Chapter 3 PRECIPITATION

3.1 Formation of Precipitation

Precipitation is the release of water from the atmosphere to reach the surface of the earth. The term "precipitation" covers all forms of water being released by the atmosphere, including snow, hail, sleet and rainfall. It is the major input of water to a river catchment area and as such needs careful assessment in any hydrological study.

The ability of air to hold water vapor is temperature dependent: the cooler the air is, the less water vapor is retained. If a body of warm, moist air is cooled, then it will become saturated with water vapor and eventually the water vapor will condense into liquid or solid water (i.e. water or ice droplets). The water will not condense spontaneously however; there need to be minute particles present in the atmosphere, called condensation nuclei, upon which the water or ice droplets form. The water or ice droplets that form on condensation nuclei are normally too small to fall to the surface as precipitation; they need to grow in order to have enough mass to overcome uplifting forces within a cloud.

Therefore, three conditions need to be met prior to precipitation forming:
(1) Cooling of the atmosphere;
(2) Condensation onto nuclei;
(3) Growth of the water/ice droplets.

3.1.1 Atmospheric Cooling

Cooling of the atmosphere may take place through several different mechanisms occurring independently or simultaneously. The most common form of cooling is from the uplift of air through the atmosphere. As air rises, the pressure decreases, which will lead to a corresponding cooling in temperature. The cooler temperature leads to less water vapor being retained by the air and conditions becoming favorable for condensation. The actual uplift of air may be caused by heating from the Earth's surface (leading to convective precipitation), an air mass being forced to rise over an obstruction such as a mountain range (this leads to orographic precipitation), or from a low pressure weather system where the air is constantly being forced upwards (this leads to cyclonic precipitation). Other mechanisms whereby the atmosphere cools include a warm air mass meeting a cooler air mass,

Chapter 3 PRECIPITATION

and the warm air meeting a cooler object such as the sea or land.

3.1.2 Condensation Nuclei

Condensation nuclei are minute particles floating in the atmosphere, which provide a surface for the water vapor to condense into liquid water upon. They are commonly less than a micron (i.e. one millionth of a meter) in diameter. There are many different substances that make condensation nuclei, including small dust particles, sea salts and smoke particles.

3.1.3 Water or Ice Droplet Growth

Water or ice droplets formed around condensation nuclei are normally too small to fall directly to the ground; that is, the forces from the upward draught within a cloud are greater than the gravitational forces pulling the microscopic droplet downwards.

In order to overcome the upward draughts, it is necessary for the droplets to grow from an initial size of 1 micron to around 3000 microns (3mm). The vapor pressure difference between a droplet and the surrounding air will cause it to grow through condensation, albeit rather slowly. When the water droplet is ice, the vapor pressure difference with the surrounding air becomes greater and the water vapor sublimates onto the ice droplet. This will create a precipitation droplet faster than condensation onto a water droplet, but is still a slow process. The main mechanism by which raindrops grow within a cloud is through collision and coalescence. Two raindrops collide and join together (coalesce) to form a larger droplet that may then collide with many more before falling towards the surface as rainfall or another form of precipitation.

Another mechanism leading to increased water droplet size is the so-called Bergeron process. The pressure exerted within the parcel of air, by having the water vapor present within it, is called the vapor pressure. The more water vapor presents, the greater the vapor pressure goes. Because there is a maximum amount of water vapor that can be held by the parcel of air, there is also a maximum vapor pressure, the so-called saturation vapor pressure. The saturation vapor pressure is greater over a water droplet than an ice droplet because it is easier for water molecules to escape from the surface of a liquid than a solid. This creates a water vapor gradient between water droplets and ice crystals so that water vapor moves from the water droplets to the ice crystals, thereby increasing the size of the ice crystals. Because clouds are usually a mixture of water vapor, water droplets and ice crystals, the Bergeron process may be a significant factor in making water droplets large enough to become rain drops (or ice/snow crystals) that overcome gravity and fall out of the clouds.

The mechanisms of droplet formation within a cloud are not completely understood. The relative proportion of condensation-formed, collision formed, and Bergeron-process-formed droplets depends very much on the individual cloud circumstances and can vary considerably.

3.1.4 Types of Precipitation

As a droplet moves around a cloud, it may freeze and thaw several times, leading to different types of precipitation (Table 3-1). Drizzle is a very light, usually uniform, precipitation consisting of numerous minute droplets with diameters in excess of 0.1mm but smaller than 0.5mm. Rain is precipitation consisting of water drops larger than 0.5mm. It can be classified as light rain when the intensity is smaller than 2.5mm/h, moderate rain when it is between 2.5 and 7.5mm/h, and heavy rain when it exceeds 7.5mm/h. Snow is precipitation in the form mainly of branched hexagonal or star-like ice crystals, resulting from direct reverse sublimation of the atmospheric water vapor; snow particles can reach the ground as single crystals, but more often than not they do so after agglomerating as snowflakes. These flakes tend to be larger at temperatures close to freezing. The specific gravity of snow can vary over a wide range, but as a rule of thumb for fresh snow, it is often taken around 0.1. Sleet is precipitation consisting of transparent pellets or grains of ice, formed because of the passage of raindrops through a layer of colder air near the ground. In British usage, the word sleet refers to precipitation consisting of melting snow or a mixture of snow and rain. Glaze or freezing rain is ice deposited by drizzle or rain on cold surfaces. Snow pellets (also called granular snow or graupel) are a form of precipitation consisting of white, opaque, small grains with diameters between roughly 0.5 and 5mm. Small hail is precipitation consisting of white, semitransparent or translucent grains with diameters ranging from 2 to 5mm. These grains are mostly round, and sometimes conical in shape, and they have a glazed appearance. Small hail falls usually accompanied by rain, when the temperature is above freezing. Soft hail consists of round, opaque grains in the same size range as small hail, but they are softer in appearance and tend to disintegrate more easily. Hail consists of balls or irregular chunks of ice with diameters between 5 and 50mm, or even larger. These lumps of ice can be transparent or they can consist of concentric layers of clear and opaque ice; such layered structure is the result of the alternating rising and falling movements during the hail formation. Hail usually falls during violent and prolonged convective storms under above-freezing temperature conditions near the ground; it can cause severe damage. Dew consists of moisture in the form of liquid drops on the ground surface and on the vegetation and other surface elements, because of direct condensation of atmospheric water vapor. It typically occurs at night on surfaces that have been cooled by outgoing long-wave radiation. Hoar frost forms in the same way as dew, but the water vapor condenses directly into ice. These ice crystals can assume a wide variety of shapes.

Classes of Precipitation Used by the UK Meteorological Office Table 3-1

Class	Definition
Drizzle	A subset of rain with droplets less than 0.5mm, intensity less than 1mm/h
Rain	Liquid water droplets between 0.5 and 7mm in diameter
Sleet	Freezing raindrops; a combination of snow and rain
Snow	Complex ice crystals agglomerated
Hail	Balls of ice between 5 and 50mm in diameter
Fog, dew and frost	Not actually precipitation; the result of interception, condensation or sublimation; can be important sources of moisture to watersheds in coastal areas and other areas subjected to persistent fog and/or clouds

Source: DAVIE T. Fundamentals of Hydrology (2nd ed). 2008.

3.2　Distribution of Precipitation

The amount of precipitation falling over a location varies both spatially and temporally (with time). The various influences on the precipitation can be divided into static and dynamic influences. Static influences are those such as altitude, aspect and slope; they do not vary between storm events. Dynamic influences are those that do change and are mostly caused by variations in the weather. At the global scale, the influences on precipitation distribution are mainly dynamic being caused by differing weather patterns, but there are static factors such as topography that can also cause major variations through a rain shadow effect. At the continental scale, large differences in rainfall can be attributed to a mixture of static and dynamic factors.

At smaller scales, the static factors are often more dominant, although it is common for quite large variations in rainfall across a small area caused by individual storm clouds to exist. For the hydrologist who is interested in rainfall at the small scale, the only way to try to characterize these dynamic variations is through having as many rain gauges as possible within a study area.

3.2.1　Static Influences on Precipitation Distribution
It is easier for the hydrologist to account for static variables such as those discussed below.

3.2.1.1　Altitude
Temperature is a critical factor in controlling the amount of water vapor that can be held by air (shown in Chapter 4). The cooler the air is, the less water vapor can be held. As temperature decreases with altitude, it is reasonable to assume that as an air parcel gains altitude, it is more likely to release the water vapor and cause higher rainfall.

This is exactly what does happen and there is a strong correlation between altitude and rainfall: so-called orographic precipitation.

3.2.1.2 Aspect

The influence of aspect is less important than altitude but it may still play an important part in the distribution of precipitation throughout a watershed. In the humid mid-latitudes (35° to 65° north or south of the equator), the predominant source of rainfall is through cyclonic weather systems arriving from the west. Slopes within a watershed that face eastwards will naturally be more sheltered from the rain than those facing westwards. The same principle applies everywhere: slopes with aspects facing away from the predominant weather patterns will receive less rainfall than their opposites.

3.2.1.3 Slope

The influence of slope is only relevant at a very small scale. Unfortunately, the measurement of rainfall occurs at a very small scale (i.e. a rain gauge). The difference between a level rain gauge on a hillslope, compared to one parallel to the slope, may be significant. It is possible to calculate this difference if it is assumed that rain falls vertically, but of course, rain does not always fall vertically. Consequently, the effect of slope on rainfall measurements is normally ignored.

Where there is a large and high land mass, it is common to find the rainfall considerably higher on one side than the other. This is through a combination of altitude, slope, aspect and dynamic weather direction influences and can occur at many different scales.

3.2.2 Forest Rainfall Partitioning

Once rain falls onto a vegetation canopy, it effectively partitions the water into separate modes of movement: throughfall, stemflow and interception loss. This is illustrated in Figure 3-1.

3.2.2.1 Throughfall

This water either falls to the ground directly, through gaps in the canopy (direct throughfall), or indirectly, having dripped off leaves, stems or branches (indirect throughfall).

The amount of direct throughfall is controlled by the canopy coverage for an area, a measure of which is the leaf area index (LAI). LAI is actually the ratio of leaf area to ground surface area and consequently has a value greater than one when there is more than one layer of leaf above the ground. When LAI is less than one, you would expect some direct throughfall to occur.

Chapter 3 PRECIPITATION

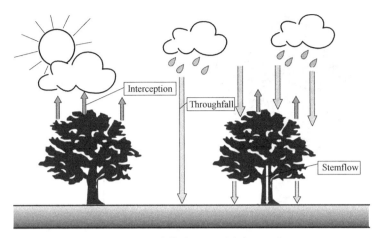

Figure 3-1 Rainfall above and below a Canopy

The amount of indirect throughfall is also controlled by the LAI, in addition to the canopy storage capacity and the rainfall characteristics. Canopy storage capacity is the volume of water that can be held by the canopy before water starts dripping as indirect throughfall. The canopy storage capacity is controlled by the size of trees, plus the area and water-holding capacity of individual leaves. Rainfall characteristics are an important control on indirect throughfall as they dictate how quickly the canopy storage capacity is filled. In reality, canopy storage capacity is a rather nebulous concept. Canopy characteristics are constantly changing and it is rare for water on a canopy to fill up completely before creating indirect throughfall. This means that indirect throughfall occurs before the amount of rainfall equals the canopy storage capacity, making it difficult to gauge exactly what the storage capacity is.

3.2.2.2 Stemflow

Stemflow is the rainfall that is intercepted by stems and branches and flows down the tree trunk into the soil. Although measurements of stemflow show that, it is a small part of the hydrological cycle (normally 2%~10% of above canopy rainfall), it can have a much more significant role. Stemflow acts like a funnel (Figure 3-2), collecting water from a large area of canopy and delivering it to the soil in a much smaller area: the surface of the trunk at the base of a tree. This is most obvious for the deciduous oak-like tree illustrated in Figure 3-2, but it still applies for other structures (e.g. conifers) where the area of stemflow entry into the soil is far smaller than the canopy watershed area for rainfall. At the base of a tree, it is possible for the water to rapidly enter the soil through flow along roots and other macropores surrounding the root structure. This can act as a rapid conduit of water sending a significant pulse into the soil water.

Figure 3-2　The Funneling Effect of a Tree Canopy on Stemflow
Source: DAVIE T. Fundamentals of Hydrology (2nd ed). 2008.

3.2.2.3　Interception

Interception is the part of precipitation that moistens the different surface elements, mainly vegetation, and is temporarily stored on them. When the surface elements are fully saturated, so that they have reached their full interception storage capacity, any excess intercepted water on them flows or drips down to the ground. In practice, the interception storage capacity is usually defined more specifically as the amount of water left on the canopy at the end of a storm, under conditions of zero evaporation and after all drip has ceased; thus during a storm the stored depth of water can exceed the storage capacity.

While water sits on the canopy, prior to indirect throughfall or stemflow, it is available for evaporation, referred to as interception loss. In some circumstances, it is possible that there is an interception gain from vegetation. When the trees are removed, there are no condensation nuclei (or far fewer) on the resultant vegetation so the water remains in the atmosphere and is "lost" in terms of water yield. Equally important is the influence of vegetation roughness. The turbulent mixing of air as wind passes over a rough canopy promotes rapid deposition of condensing water. The overall result of this is that the removal of trees leads to less water in the river.

3.3　Measurement of Precipitation

For hydrological analysis, it is important to know how much precipitation has fallen and when this occurred. The usual expression of precipitation is as a vertical depth of liquid water. Rainfall is measured by millimeters or inches depth, rather than by volume such as liters or cubic meters. The measurement is the depth of water that would accumulate on the surface if all the rain remained where it had fallen. Snowfall may also be expressed as a depth, although for hydrological purposes it is most usefully described in water equivalent depth (i. e. the depth of water that would be present if the snow melted). This is a recognition that snow takes up a greater volume (as much as 10% more) for the same amount of liquid water.

It is a strong argument that there is no such thing as precipitation measurement at the watershed scale as it varies so tremendously over a small area. The logical endpoint to this argument is that all measurement techniques are in fact precipitation estimation techniques. For the sake of clarity in this text, precipitation measurement techniques refer to the methods used to quantify the volume of water present, as opposed to estimation techniques where another variable is used as a surrogate for the water volume.

3.3.1 Direct Precipitation Measurement

3.3.1.1 Rainfall Measurement

The instrument for measuring rainfall is called a rain gauge. A rain gauge measures the volume of water that falls onto a horizontal surface delineated by the rain gauge rim (Figure 3-3). The volume is converted into a rainfall depth through division by the rain gauge surface area. The design of a rain gauge is not as simple as it may seem at first glance, as there are many errors and inaccuracies that need to be minimized or eliminated. Table 3-2 shows the rain-gauge density for various types of areas.

Figure 3-3　A Rain Gauge Sitting above the Surface to Avoid Splash
Source: DAVIE T. Fundamentals of Hydrology (2nd ed). 2008.

Rain-gauge Density　　　　　　　　　　Table 3-2

Area	Rain-gauge Density
Plains	1 for 520km²
Elevated Area	1 for 260km² to 390km²
Hilly and very Heavy Rainfall Areas	1 for 130km² preferably with 10% of the rain-gauge stations equipped with the self-recording type

Source: RAO S N. Hydrogeology: Problems with Solutions. 2016.

The accuracy and errors involved in measuring rainfall were studied in a large number of scientific literature. It needs to be borne in mind that a rain gauge represents a very small point measurement (or sample) from a much larger area that is covered by the rainfall. Any errors in measurement can be amplified hugely because the rain gauge collection area represents such a small sample size. Because of this amplification, it is extremely important that the design of a rain gauge negates any errors and inaccuracies.

There are four main sources of error in measuring rainfall that need consideration in designing a method for the accurate measurement of rainfall:

① Losses due to evaporation;
② Losses due to wetting of the gauge;
③ Over-measurement due to splash from the surrounding area;
④ Under-measurement due to turbulence around the gauge.

(1) Siting of a Rain Gauge

Once the best measurement device has been chosen for a location, there is still a considerable measurement error that can occur through incorrect siting. The major problem of rain gauge siting in hydrology is that the scientist is trying to measure the rainfall at a location that is representative of a far greater area. It is extremely important that the measurement location is an appropriate surrogate for the larger area. If the area of interest is a forested catchment, it is reasonable to place your rain gauge beneath the forest canopy; likewise, within an urban environment it is reasonable to expect interference from buildings because this is what is happening over the larger area. What is extremely important is that there are enough rain gauges to try to quantify the spatial and temporal variations.

The rule-of-thumb method for siting a rain gauge is that the angle when drawn from the top of the rain gauge to the top of the obstacle is less than 30° (Figure 3-4). This can be estimated as at least twice the height of the obstacle away from the gauge. Extra care is needed to allow for the future growth of trees, so that the distance apart is always at least twice the height of an obstacle during the rainfall event.

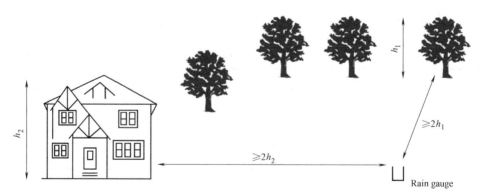

Figure 3-4　Siting of a Rain Gauge away from Obstructions

(2) Gauges for the Continuous Measurement of Rainfall

The standard rain gauge collects water beneath its funnel and this volume is read once a day. Often in hydrology, the data needs to be measured at a finer timescale than this, particularly in the case of individual storms, which often last much less than a day. The most common modern method for measuring continuous rainfall uses a tipping-bucket rain gauge. These simple devices can be installed relatively cheaply, although they do require a data-logging device nearby. The principle behind the tipping bucket rain gauge is that as the rain falls it fills up a small "bucket" that is attached to another "bucket" on a balanced cross arm (Figure 3-5). The "buckets" are very small plastic containers at the end of each cross arm. When the bucket is full, it tips the balance so that the full bucket leans down and empties out. At the time of tipping, a magnet attached to the balance arm closes a small reed switch, which sends an electrical signal to a

Figure 3-5 The Insides of a Tipping-bucket Rain Gauge

data-logging device. This then records the exact time of the tipped bucket. If the rain continues to fall, it fills the bucket on the other end of the cross arm until it tips the balance arm, too, sending another electrical impulse to the data logger. In this way, a near continuous measurement of rainfall with time can be obtained.

It is important that the correct size of tipping bucket to be used for the prevailing conditions. If the buckets are too small, then a very heavy rainfall event will cause them to fill too quickly and water will be lost through overspill while the mechanism tips. If the buckets are too large then a small rainfall event may not cause the cross arm to tip and the subsequent rainfall event will appear larger than it actually is.

3.3.1.2 Snowfall Measurement

The measurement of snowfall has similar problems to those presented by rainfall, but they are often more extreme. There are two methods used for measuring snowfall: using a gauge like a rain gauge; or measuring the depth that is present on the ground. Both of these methods have very large errors associated with them, predominantly caused by the way that snow falls through the atmosphere and is deposited on the gauge or ground. Mostly, although not all, snowflakes are more easily transported by the wind than raindrops. When the snow reaches the ground, it drifts. This can be contrasted to liquid water where, upon reaching the ground, it is either absorbed by the soil or moves across the surface. Rainfall is very rarely picked up by the wind again and redistributed in the manner that drifting snow is. For the snow gauge, this presents problems that are analogous to rain splash. For the depth gauge, the problem is due to uneven distribution of the snow surface: it is likely to be deeper in certain situations.

3.3.1.3 Forest Rainfall Measurement

The most common method of assessing the amount of canopy interception loss is to measure the precipitation above and below a canopy and assume that the difference is from interception.

(1) Above-canopy Precipitation

To measure above-canopy precipitation, a rain gauge may be placed on a tower above the canopy. The usual rain gauge errors apply here, but especially the exposure to the wind. The top of a forest canopy tends to be rough and is very good for allowing turbulent transfer of evaporated water. The turbulent air is not so good for measuring rainfall. An additional problem for any long-term study is that the canopy is not static; the tower needs to be raised every year so that it remains above the growing canopy.

One way around the tower problem is to place a rain gauge in a nearby clearing and assume that what falls there is the same amount as directly above the canopy nearby. This is often perfectly reasonable to assume, particularly for long-term totals, but care must be taken to ensure the clearing is large enough to avoid obstruction from nearby trees (Figure 3-4).

(2) Throughfall

Throughfall is the hardest part of the forest hydrological cycle to measure. This is because a forest canopy is normally variable in density and therefore throughfall is spatially heterogeneous. One common method is to place numerous rain gauges on the forest floor in a random manner. If you are interested in a long-term study, then it is reasonable to keep the throughfall gauges in fixed positions. However, if the study is investigating individual storm events, the best practice is considered to move the gauges to new random positions between storm events. In this way, the throughfall catch should not be influenced by gauge position. To derive an average throughfall figure, it is necessary to come up with a spatial average in the same manner as for areal rainfall estimates.

To overcome the difficulty of a small sampling area (rain gauge) measuring something notoriously variable (throughfall), some investigators have used either troughs or plastic sheeting. Troughs collect over a greater area and have proved to be very effective. Plastic sheeting is the ultimate way of collecting throughfall over a large area, but has several inherent difficulties. The first is purely logistical in that it is difficult to install and maintain, particularly to make sure there are no rips. The second is that by having an impervious layer above the ground, there is no, or very little, water entering the soil. This might not be a problem for a short-term study but will be over the longer term, especially if the investigator is interested in the total water budget, it may also place the trees under stress through lack of water, thus leading to an altered canopy.

(3) Stemflow

The normal method of measuring stemflow is to place collars around a tree trunk that capture all the water flowing down the trunk. On trees with smooth bark, this may be relatively simple but is very difficult on rough bark such as found on many conifers. It is important that the collars be sealed to the tree so that no water can flow underneath and that they are large enough to hold all the water flowing down the trunk. The collars should be sloped to one side so that the water can be collected or measured in a tipping bucket rain gauge.

3.3.2 Surrogate Precipitation Measurements

The direct estimation of areal precipitation is an attractive proposition. There are two techniques that make some claim to achieving this: radar and satellite remote sensing. These approaches have many similarities, but they differ fundamentally in the direction of measurement. Radar looks from the Earth up into the atmosphere and tries to estimate the amount of precipitation falling over an area. Satellite remote sensing looks from space down towards the earth surface and attempts to estimate the amount of precipitation falling over an area.

3.3.2.1 Radar

The main use of ground-based radar is in weather forecasting where it is applied to track the movement of rain clouds and fronts across the Earth's surface. There are several techniques used for radar, although they are all based on similar principles. Radar is an acronym (radio detection and ranging). A wave of electromagnetic energy is emitted from a unit on the ground, and the amount of wave reflection and return time are recorded. The more water there is in a cloud, the more electromagnetic energy is reflected back to the ground and detected by the radar unit. The quicker the reflected wave reaches back to the ground, the closer the cloud is to the surface. The most difficult part of this technique is in finding the best wavelength of electromagnetic radiation to emit and detect. It is important that the electromagnetic wave be reflected by liquid water in the cloud, but not atmospheric gases and/or changing densities of the atmosphere. A considerable amount of research effort has gone into trying to find the best wavelengths for ground-based radar to use. Studies have shown a good correlation between reflected electromagnetic waves and rainfall intensity. Therefore, this can be considered as a surrogate measure for estimating rainfall. If an accurate estimate of rainfall intensity is required then a relationship has to be derived using several calibrating rain gauges. Herein lies a major problem: with this type of technique, there is no universal relationship that can be used to derive rainfall intensity from cloud reflectivity. An individual calibration has to be derived for each site and this may involve several years of measuring point rainfall coincidentally with cloud reflectivity. This is not a cheap option and the cost prohibits its widespread usage, particularly in areas with poor rain gauge coverage.

3.3.2.2 Satellite Remote Sensing

The atmosphere-down approach of satellite remote sensing is quite different from the ground-up approach of radar. That is because that the sensor is looking at the top of a cloud rather than the bottom. It is well established that a cloud most likely to produce rain has an extremely bright and cold top. These characteristics can be observed from space by a satellite sensor.

The most common form of satellite sensor is passive (this means it receives radiation from another source, normally the sun, rather than emitting any itself as the way radar does) and detects radiation in the visible and infrared wavebands. LANDSAT, SPOT and AVHRR are examples of satellite platforms of this type. By sensing in the visible and infrared part of the electromagnetic spectrum, the cloud brightness (visible) and temperature (thermal infrared) can be detected. This so-called "brightness temperature" can then be related to rainfall intensity via calibration with point rainfall measurements, in a similar fashion to ground-based radar. One of the problems with this approach is that it is sometimes difficult to distinguish between snow reflecting light on the ground and clouds reflecting light in the atmosphere, which have similar brightness temperature values.

Another form of satellite sensor that can be used is passive microwave. The earth emits microwaves (at a low level) that can be detected from space. When there is liquid water between the Earth's surface and the satellite sensor (i.e. a cloud in the atmosphere), some of the microwaves are absorbed by the water. A satellite sensor can therefore detect the presence of clouds (or other bodies of water on the surface) as a lack of microwaves reaching the sensor.

Although there was some success in the method, it is at a scale of little use to watershed scale hydrology as the best resolution available is around $10km \times 10km$ grid sizes. Satellite remote sensing provides an indirect estimate of precipitation over an area but is still a long way from operational use. Studies have shown that it is an effective tool where there is poor rain gauge coverage, but in countries or regions with high rain gauge density, it does not improve the accuracy of estimation of areal precipitation. What is encouraging about the technique is that nearly all the world is covered by satellite imagery so that it can be used in sparsely gauged areas. The new generation of satellite platforms being launched in the early 21st century will have multiple sensors on them so it is feasible that they will be measuring visible, infrared and microwave wavebands simultaneously. This will improve the accuracy considerably but it must be borne in mind that it is an indirect measure of precipitation and will still require calibration to a rain gauge set.

3.4 Estimation of Spatially Distributed Precipitation

In hydrology, we typically want to estimate the total volume of water delivered to a given area (e.g. a watershed) within some period of time. That is, we want to know the average precipitation depth over the area, i.e.,

$$P = \frac{1}{A}\iint_A p(x,y)\,dx\,dy \tag{3-1}$$

where P——the average depth of precipitation over the watershed;

A——the watershed area;

$p(x,y)$——the spatial distribution of cumulative precipitation for the time period of interest.

Once P is determined, the total volume of water delivered to the area is simply as PA. The measurement techniques described here have all concentrated on measuring rainfall at a precise location (or at least over an extremely small area). In fact, the hydrologist needs to know how much precipitation has fallen over a far larger area, usually a watershed (shown in Chapter 5). To move from point measurements to a spatially distributed estimation, it is necessary to employ some form of spatial averaging. The spatial averaging must attempt to account for an uneven spread of rain gauges in the watershed and the various factors that we know influence rainfall distribution (e.g. altitude, aspect and slope).

A simple arithmetic mean will only work where a watershed is sampled by uniformly spaced rain gauges and where there is no diversity in topography. It is very rare to use this technique. There are different statistical techniques that address the spatial distribution issues, and with the growth in use of Geographic Information Systems (GIS), it is often a relatively trivial matter to do the calculation. As with any computational task, it is important to have a good knowledge of how the technique works so that any shortcomings are fully understood. Three techniques are described here: Thiessen's polygons, the hypsometric method and the isohyetal method.

3.4.1 Thiessen's Polygons

Thiessen was an American engineer working around the start of the 20th century who devised a simple method of overcoming an uneven distribution of rain gauges within a watershed (very much the norm). Essentially Thiessen's polygons attach a representative area to each rain gauge. The size of the representative area (a polygon) is based on how close each gauge is to the others surrounding it.

In Figure 3-6, the area of each polygon is denoted as a_i. Locations of rain gauges are indicated by bullet points.

Each polygon is drawn on a map; the boundaries of the polygons are equidistant from each gauge and drawn at a right angle (orthogonal) to an imaginary line between two gauges (Figure 3-6). Once the polygons have been drawn, the area of each polygon surrounding a rain gauge is found. The spatially averaged rainfall (R) is calculated using Equation (3-2):

$$R = \sum_{i=1}^{n} \frac{r_i a_i}{A} \qquad (3-2)$$

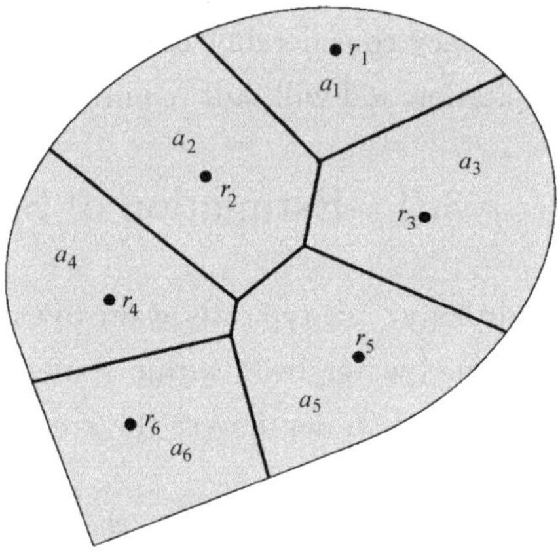

Figure 3-6 Thiessen's Polygons within an Imaginary Watershed

Note: Locations of rain gauges are indicated by bullet points.

where r_i——the rainfall at gauge i;

a_i——the area of the polygon surrounding rain gauge i;

A——the total watershed area.

The areal rainfall value using Thiessen's polygons is a weighted mean, with the weighting being based upon the size of each representative area (polygon). This technique is only truly valid where the topography is uniform within each polygon, so that it can be assumed that the rainfall distribution is uniform within the polygon. This would suggest that it could only work where the rain gauges are located initially with this technique in mind.

3.4.2 Hypsometric Method

Since it is well known that rainfall is positively influenced by altitude (i.e. the higher the altitude, the greater the rainfall), it is reasonable to assume that knowledge of the watershed elevation can be brought to bear on the spatially distributed rainfall estimation problem. The simplest indicator of the watershed elevation is the hypsometric (or hypsographic) curve. This is a graph showing the proportion of a watershed above or below a certain elevation (Figure 3-7). The values for the curve can be derived from maps using a plan meter or using a digital elevation model (DEM) in a GIS.

The hypsometric method of calculating spatially distributed rainfall then calculates a weighted average based on the proportion of the watershed between two elevations and the measured rainfall between those elevations (Equation 3-3).

$$R = \sum_{j=1}^{m} r_j p_j \tag{3-3}$$

where r_j——the average rainfall between two contour intervals;

p_j——the proportion of the total watershed area between those contours (derived from the hypsometric curve).

The r_j value may be an average of several rain gauges where there is more than one at a certain contour interval. This is illustrated in Figure 3-7 where the shaded area (p_3) has two gauges within it. In this case, the r_j value will be an average of r_4 and r_5. Intuitively this is producing representative areas for one or more gauges based on contours and spacing, rather than just on the latter as for Thiessen's polygons. There is an inherent assumption that elevation is the only topographic parameter affecting rainfall distribution (i.e. slope and aspect are ignored). It also assumes that the relationship between altitude and rainfall is linear, which is not always the case and warrants exploration before using this technique.

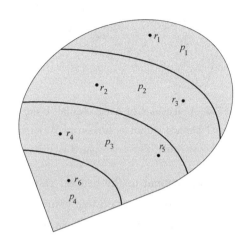

Figure 3-7 Calculation of Areal Rainfall Using the Hypsometric Method within an Imaginary Watershed

Note: The shaded region is between two contours. In this case, the rainfall is an average value between the two gauges within the shaded area. Bullet points indicate locations of rain gauges.

3.4.3 Isohyetal and Other Smoothed Surface Techniques

Where there is a large number of gauges within a watershed, the most obvious weighting to apply on a mean is based on measured rainfall distribution rather than on surrogate measures as described above. In this case, a map of the watershed rainfall distribution can be drawn by interpolating between the rainfall values, creating a smoothed rainfall surface. The traditional isohyetal method involves drawing isohyets (lines of equal rainfall) on the map and calculating the area between each isohyet. The spatial average could then be calculated by Equation (3-4)

$$R = \frac{1}{A} \sum_{i=1}^{n} a_i r_i \tag{3-4}$$

where a_i——the area between each isohyet;

r_i——the average rainfall between the isohyets.

This technique is analogous to hypsometric method, except in this case the contours will

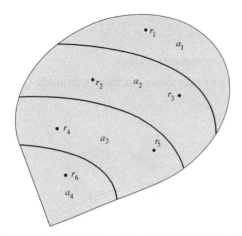

Figure 3-8 Calculation of Areal Rainfall Using the Isohyetal Method within an Imaginary Watershed
Note: Bullet points indicate locations of rain gauges.

be of rainfall rather than elevation (Figure 3-8).

With the advent of GIS, the interpolating and drawing of isohyets can be done relatively easily, although there are several different ways of carrying out the interpolation. The interpolation subdivides the watershed into small grid cells and then assigns a rainfall value for each grid cell (this is the smoothed rainfall surface). The simplest method of interpolation is to use a nearest neighbor analysis, where the assigned rainfall value for a grid square is proportional to the nearest rain gauges. A more complicated technique is to use Kriging, where the interpolated value for each cell is derived with knowledge on how closely related the nearby gauges are to each other in terms of their covariance.

An additional piece of information that can be obtained from interpolated rainfall surfaces is the likely rainfall at a particular point within the watershed. This may be more useful information than total rainfall over an area, particularly when needed for numerical simulation of hydrological processes.

The difficulty in moving from the point measurement to a spatially distributed average is a prime example of the problem of scale that besets hydrology. The scale of measurement (i. e. the rain gauge surface area) is far smaller than the watershed area that is frequently our concern. Is it feasible to simply scale up our measurement from point sources to the overall watershed? Alternatively, should there be some form of scaling factor to acknowledge the large discrepancy? There is no easy answer to these questions and they are the type of problems that research in hydrology will be investigating later in the 21st century.

3.5 Temporal Characteristics of Precipitation

Water depth is not the only rainfall measure of interest in hydrology; also of importance is the rainfall intensity and storm duration. These are simple to obtain from an analysis of rainfall records using frequency analysis. The rainfall needs to be recorded at a short time interval (i. e. an hour or less) to provide meaningful data.

Chapter 3　PRECIPITATION

One minute it may be raining heavily, while a few minutes later it may not be raining at all. Our ability to forecast this temporal variation even a few hours in advance is limited and our ability to forecast several days in advance is almost impossible. The extreme uncertainty associated with precipitation forecasts suggests substantial randomness in the occurrence of precipitation at a point in space and implies the necessity of a probabilistic approach for characterizing the temporal variations in precipitation.

As noted previously, if we examine a typical short-term hyetograph, we see that precipitation is organized into discrete (storm) events of varying amounts (storm depths). Average precipitation intensity is the rate of precipitation over a specified time period, the precipitation depth divided by the time over which that depth is recorded.

For example, the data in Figure 3-9 is reported for each 10-minute (i. e. the hourly precipitation intensity). The hourly rainfall intensity for the period shown varies from 0.5 to about 6.5 in/h. Average rainfall intensities depend on the time period over which the computation is done.

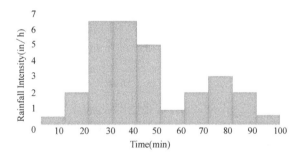

Figure 3-9　A Hyetograph for a Rainfall Event on some Site in Louisville, Kentucky, Showing Temporal Variability in Precipitation

That is, the variation in hourly rainfall intensity typically will be much greater than the variation in 6-hour intensity (average over a longer time), but less than that of 15-minute intensity (average over a shorter time). Average precipitation intensity for a storm is the total depth of precipitation for the storm divided by the storm length. In general, the longer the storm-event duration is, the less the (average) storm intensity will be. However, the greater the storm duration is, the greater the storm depth will be.

Figure 3-10 is an example of the rainfall intensity for a rain gauge at Bradwell-on-Sea, Essex, UK. It is evident from the diagram that the majority of rain falls at very low intensity: 0.4mm/h is considered as light rain. This may be misleading as the rain gauge recorded rainfall every hour and the small amount of rain may have fallen during a shorter period than an hour, i. e. a higher intensity but lasting for less than an hour. During the period of measurement, there were recorded rainfall intensities greater than 4.4mm/h, but they were so few as to not show up on the histogram scale used in Figure 3-10. This may be a reflection of only two years of records being analyzed, which introduces an extremely important concept in hydrology: the frequency-magnitude relationship. If we think of the relative frequency as a probability then we can say that the chances of having a low rainfall event are very high: a low magnitude-high frequency event. Conversely, the chances of having a rainfall intensity greater than 5mm/h are very low (but not impossible): a high magnitude-low frequency event.

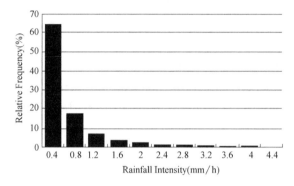

Figure 3-10   Rainfall Intensity Curve (Data are hourly-recorded rainfall from April 1995 to April 1997)

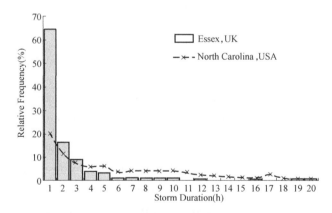

Figure 3-11   Storm Duration Curves

Note: The bars are for the same data set as Figure 3-10 and the broken line is for Ahoskie, North Carolina.

In Figure 3-11, the storm duration records for two different sites are compared. The Bradwell-on-Sea site has the majority of its rain events lasting one hour or less. In contrast, the Ahoskie site has only 20% of its storms lasting one hour or less but many more than Bradwell-on-Sea that last four hours or more. When the UK site rainfall and intensity curves are looked at together (i.e. Figure 3-10 and Figure 3-11), it can be stated Brad-

Chapter 3　PRECIPITATION

well-on-Sea experiences a predominance of low intensity, short duration rainfall events and very few long duration, high intensity storms. This type of information is extremely useful to a hydrologist investigating the likely runoff response that might be expected for the rainfall regime.

Summary

Precipitation is the main input of water within a watershed. Its measurement is fraught with difficulties and any small errors will be magnified enormously at the watershed scale. It is also highly variable in time and space. Despite these difficulties, precipitation is one of the most regularly measured hydrological variables, and good rainfall records are available for many regions in the world. A forest canopy partitions rainfall into components that move at different rates towards the soil surface. The nature of the canopy (leaf size distribution and leaf area index) determines the impact that a canopy has on the water balance equation. Analysis of rainfall can be carried out with respect to trying to find a spatial average or looking at the intensity and duration of storm events. Although there are techniques available for estimating precipitation, their accuracy is not such that it is superior to a good network of precipitation gauges.

Translation of Some Sections

部分章节参考译文

3.1　降雨的形成

降水是水从大气释放到地球表面的过程。"降水"一词包括大气所释放的所有形式的水，包括雪、冰雹、雨夹雪和雨。它是河流的主要输入源，因此，在水文研究中有重要地位。

空气持有水蒸气的能力取决于温度，空气越冷，其所持有的水蒸气越少。温暖潮湿的空气受冷后，水蒸气逐渐饱和，最终使水蒸气凝结成液体或固体（即水滴或冰晶）。然而，水并不会自发凝结，大气中需要有凝结核（即微小的粒子），水或冰晶在凝结核上形成。则一开始，在凝结核上形成的水滴或冰晶通常太小，不能像降水一样落到地面；随着水滴或冰晶的增大，当增大到具有足够的质量来克服云中的上升力时，其就会降落到地面。

因此，在降水形成之前，需要满足三个条件：
（1）大气冷却；
（2）凝结核；
（3）水滴/冰晶的增大。

3.1.1　大气冷却

大气冷却可以由几个不同的机制单独或同时导致。最常见的冷却方式是空气上升。空

41

气上升和压力降低，导致相应的温度下降。较低的温度使空气中的水蒸气易于冷凝。空气的上升可能是由于地表的加热（易导致对流性降水），气团被迫沿着如山脉这样的障碍物上升（易导致地形降水）；或迫使空气不断上升的低压天气系统（易导致气旋性降水）。与较冷空气团相遇的暖空气团和与较冷物体（如海洋或陆地）相遇的暖空气团也会导致大气冷却。

3.1.2 凝结核

凝结核是漂浮在大气中的微小粒子，它为水蒸气凝结成液态水提供了一个基面。凝结核的直径通常小于$1\mu m$。小尘埃颗粒、海盐和烟雾颗粒等均可形成凝结核。

3.1.3 小水滴/冰晶增大

由于最初凝结核周围形成的水滴/冰晶太小，通常不能直接落到地面。也就是说，云中气流对微小水滴/冰晶向上拉的力大于将其向下拉的重力。

为了克服向上的拉力，水滴必须从起初的$1\mu m$增长到$3mm$左右。尽管水滴增长速度相当缓慢，但由于其与周围空气之间存在蒸气压差，这个过程会持续进行。当水滴凝结成冰时，因其与周围空气的蒸气压差变得更大，水蒸气会升华到冰晶上。相比凝结到水滴上，这个过程更快，但都仍是一个缓慢的过程。水滴在云中生长的主要机制是碰撞与合并，即两个水滴碰撞并结合在一起（聚结）形成一个更大的水滴，再与更多的水滴碰撞，然后以降雨或其他形式的降水落向地表。

另一个导致水滴/冰晶尺寸增大的机制是贝吉龙过程。液体（或固体）的表面存在着该物质的蒸汽，这些蒸汽对液体表面产生的压强就是该液体的蒸气压。蒸汽越多，蒸气压就越大。因为可以被空气包裹的水蒸气有一个最大量，因此也有一个最大的蒸气压，即饱和蒸气压。相对于固体，水分子更容易从液体表面逸出，所以水滴的饱和蒸气压比冰晶的大。这是因为水滴和冰晶之间形成了蒸气压梯度，使水蒸气从水滴移动到冰晶，从而增大了冰晶的尺寸。由于云通常是水蒸气、水滴和冰晶的混合物，贝吉龙过程很可能是使水滴/冰晶大到足以克服重力并从云中落下的一个重要因素。

云中水滴/冰晶的形成机制尚不明确。凝结形成、碰撞形成和贝吉龙过程形成的水滴/冰晶的相对比例在很大程度上取决于云周围的环境，并且可以有很大的不同。

3.2 降水分布

降水量随空间和时间而变化。降水的影响因素分为静态因素和动态因素。静态因素是指海拔、坡向和坡度等在不同降雨事件中的不变因素。动态因素主要是指由天气变化引起的那些不断变动的因素。在全球范围内，对降水分布的影响主要来自由不同天气模式导致的动态变化，也包含因雨影效应引起重大变化的静态因素（如地形）。在陆地范围内，降雨量的巨大差异可归因于静态和动态因素的混合。

在更小范围内，尽管由单个风暴云引起的小范围降雨很常见，但静态因素往往占主导地位。对于那些对小范围降雨感兴趣的水文学家来说，试图描述这些动态变化的唯一方法是在研究区内设置尽可能多的雨量计。

3.2.1　影响降水分布的静态因素

以下这些静态因素是较易描述的：

3.2.1.1　海拔高度

温度是控制空气中水气含量的一个关键因素（见第 4 章）。气温越低，空气所持有的水蒸气就越少。气温是随着海拔升高而降低的，因此我们可以假设，空气在上升过程中释放出水气，并导致降水量增大。

海拔高度和降水量之间有着强相关性，即是所谓的地形降水。

3.2.1.2　地势

相比海拔高度，地势对降水分布的影响要小一些，但其对整个流域降水分布仍然起着重要的作用。在潮湿的中纬度地区（赤道以北或以南纬度 35°～65°），主要的降水来源于西部的气旋性天气。因此，面向东方的斜坡比面向西方的斜坡更容易受雨水的冲击。有一个普遍的规律是：背风坡比迎风坡的降水量小。

3.2.1.3　坡度

坡度只在很小的范围影响降水。同时，降雨量是在很小的范围内进行测量的（即利用雨量计）。安置在流域内某山坡上的雨量计与坡下的雨量计所测的降雨量可能会有很大差异。假设雨是垂直下降的，这种差异是可以通过计算得出的。然而实际上，雨并不总是垂直下降的。因此，人们通常选择忽略坡度对降雨测量的影响。

通常，在有大片高地的地区，其一侧的降雨量比另一侧要大得多。这是海拔高度、坡度、坡向和动态天气综合作用的结果，在不同的尺度上均可发生。

3.2.2　森林降雨分区

一旦雨水落在植被冠层上，其将进入不同的状态：净雨、茎流和截留，如图 3-1 所示。

图 3-1　冠层上下的雨量

3.2.2.1　净雨

净雨是指直接从林冠的缝隙流到地面的那部分水（直接净雨），或者是间接地从叶子、茎或树枝上滴落下来的那部分水（间接净雨）。

直接净雨量由林冠覆盖率控制，可由叶面积指数（LAI）反映。LAI是叶面积与地表面积的比值。当地面上有一层以上的落叶时，LAI的值大于1。当LAI小于1时，一部分雨可直接穿透到达地面。

间接净雨量受LAI、林冠蓄水量和降雨特性共同控制。林冠蓄水量是指可在水开始滴落前被林冠作为间接贯穿物固定的水量，其与树木大小、单叶面积和持水量有关。降雨特性是间接净雨量的一个重要控制因素，因为它们决定了林冠蓄水的速度。实际上，林冠蓄水量是一个模糊的概念。林冠的特性是不断变化的，在形成间接净雨之前，林冠上很难能完全填满水。这意味着，在降雨量等于林冠蓄水量之前，会发生间接净雨，因此很难准确确定蓄水量是多少。

3.2.2.2 茎流

茎流是指由茎或枝截留并沿树干流入土壤的那部分降雨。测量结果表明，茎流虽然只占水文循环的一小部分（通常占冠层以上降雨量的2%~10%），但它可以发挥重要的作用。茎流就像一个漏斗（图3-2），从大面积的树冠中收集水分，将其输送到位于树干底部的土壤中。如图3-2所示，落叶栎树最为明显，茎流也适用于茎流进入土壤的面积远小于林冠集水区面积的其他树木（例如针叶树）。在树的根部，水有可能沿着树根和围绕树根结构的大孔隙快速进入土壤，这很像输送脉冲的快速通道。

图3-2 林冠对茎流的漏斗效应

资料来源：DAVIE T. Fundamentals of Hydrology (2nd ed). 2008.

3.2.2.3 截留

截留是指润湿表面（主要是植被）并暂存其上的那部分降水。当地面完全饱和，即达到其全部截留储存容量时，其上任何多余的截留水都会流向或滴落到地面。实际上，截留蓄水量通常被具体地定义为：暴雨结束时，在零蒸发条件下，所有滴水停止后留在冠层上的水量。因此，在降水期间，蓄水量可能会超过其截留蓄水量。

当降水落到林冠上形成净雨和茎流之前，会蒸发一部分，蒸发掉的这部分称为截留损失。一般情况下，植被截留是有利于人类生存的。当植被受到破坏时，其不产生凝结核（或减少凝结核产生量），因此降雨量会减少，造成水量方面的"流失"。植被粗糙度也会影响降雨的形成。当风通过粗糙的林冠时，空气的湍流混合促进冷凝水的快速沉积。因此，森林砍伐可导致河水水量减少。

3.3 降水量的测量

了解降水量和降水时间对水文分析很重要。降水量通常表示为液态水的垂直深度。降雨量是以毫米（mm）或英寸（inch）来计量的，而不是以体积（L, m³）来计量的。降雨量测量的是所有的雨都留在原来地方时地表上积聚水的深度。降雪也可以用深度表示，但在水文学中，最常用的描述是水当量深度（即雪融化后的水深）。与相同质量的液态水相比，雪所占的体积更大（高达10％以上）。

3.3.1 直接降水量的测量
3.3.1.1 降雨量测量

测量降雨量的仪器叫做雨量计。雨量计测量的是落在由雨量计边缘划定的水平面上的水量（图3-3）。通过体积除以雨量计表面积转换成降雨深度。雨量计在设计时需要将错误和误差降至最低或消除，所以雨量计的设计并不那么简单。表3-2显示了不同类型区域的雨量计建议安置密度。

雨量计的建议安置密度　　　　表3-2

区域类型	雨量计密度
平原	每520km² 安置1个雨量计
高地	每260km²～390km² 安置1个雨量计
山区或多雨地区	每130km² 安置一个雨量计，其中10％的雨量计最好具有自动记录功能

资料来源：RAO S N. Hydrogeology: Problems with Solutions. 2016.

大量文献中记载了有关雨量计测量降雨量的准确性和误差方面的研究。需要注意的是，雨量计所测的降雨量值仅代表被降雨覆盖的区域中一个非常小的点（或样本）。因此，任何测量误差都会被放大。由于这种放大作用，雨量计的安置必须减少误差和错误。

在设计精确测量降雨量的方法时，有4个主要误差源需要考虑：
① 蒸发损失；
② 仪表受潮造成的损失；
③ 周围雨滴飞溅导致测量数值过高；
④ 仪表周围的湍流导致测量数值过低。

（1）雨量计的选址

为某测量点选择了最佳的测量设备后，仍会存在由于不正确的选址而导致的较大测量误差。雨量计应选址在一个能代表更大区域的位置来测量降雨量。如果测量区域是一片森林，雨量计应放置在森林冠层下；如果测量区域在城市中，雨量计的测量值很大程度上会受到建

图3-3　安装在地面上的雨量计

筑物的干扰，因此，要有足够数量的雨量计。

雨量计选址的经验法则是：从雨量计顶部到障碍物顶部的角度小于30°（图3-4）。也就是说，障碍物离雨量计的距离至少应是其高度的两倍。同时，还需要考虑到树木的生长。

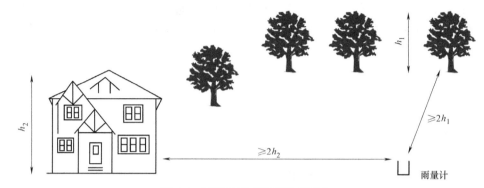

图 3-4　雨量计远离障碍物的距离

图 3-5　翻斗式雨量计的内部结构

（2）可连续测量的雨量计

标准的雨量计收集雨水并每天读取一次数值。但通常暴雨持续不到一天，因此暴雨数据的测量需要在比这更精细的时间尺度上进行。最常用的连续降雨测量方法是翻斗式雨量计。翻斗式雨量计的原理是：当雨落下时，它会填满一个小"桶"，这个小"桶"与平衡横臂上的另一个"桶"相连（图3-5）。"桶"是每个横臂末端非常小的塑料容器。当"桶"装满时，它会使天平倾斜，装满的水桶下降并使水倾出。在倾斜时，与平衡臂相连的磁铁通过闭合小簧片开关向数据记录设备发送电脉冲信号，记录倒桶的准确时间。如果降雨一直持续，它会填满横臂另一端的水桶，直到平衡臂倾斜至另一端，从而向数据记录器发送另一个电脉冲。通过这种方法，可以获得几乎连续的瞬时降雨量。

桶的尺寸非常重要。如果桶太小，强降雨会使它们填充得太快，装置倾斜时会造成一部分水损失。如果桶太大，小的降雨事件甚至可能不会引起横臂倾斜，导致随后的降雨事件会因桶中残存的水而使其测量值大于实际情况。

3.3.1.2　降雪的测量

降雪量的测量会遇到与降雨相似的问题，且往往更极端。测量降雪量的方法有两种：一是像雨量计一样使用量具；二是测量其在地面的深度。由于降雪通过大气层并沉积在仪表或地面上，因此这两种方法都存在很大的误差。雪花比雨滴更容易被风吹走。当雪到达地面时，很容易漂移。这与液态水形成鲜明的对比：液态水到达地面后，要么被土壤吸收，要么在地表移动。降雨很少再被风卷起，漂移后再分配。对于测雪仪来说，也存在类似于雨点飞溅的问题。而深度计的误差主要来源于积雪的不均匀分布。

3.3.1.3 森林降雨量的测量

评估林冠截留损失量的最常用方法是测量林冠上下的降水量，并假设其差异由截留量导致。

(1) 冠层以上降水量

为了测量林冠上方的降水量，需要将雨量计置于在林冠上方的塔架上。一般的雨量计可以克服风引起的误差。森林冠层的顶部往往很粗糙，虽有利于蒸发水的湍流传输，但湍流的空气并不利于降雨量的测量。对于长期研究来说，考虑到林冠的生长，塔架每年都需要升高，以便它保持在林冠之上。

另外一种方法是在森林附近清理出一片空地，在空地上放置一个雨量计，并假设此雨量计与附近冠层的正上方的雨量计的作用相同。对长期降雨总量的测量，这是完全合理的替代方案。但必须确保空地足够大，以避免附近树木对测量过程的影响（图3-4）。

(2) 净雨量

净雨量是森林水文循环中最难测量的部分。因为森林冠层的密度是变量，因此净雨在空间分布上是不均匀的。常见的方法是在森林地面随机放置许多雨量计。如果是进行长期研究，可将净雨雨量计保持在固定位置。如果研究单个暴雨事件，应在暴雨事件发生之间随机移动雨量计，来确保所测的降雨量不受仪表位置的影响。为得到净雨平均值，有必要以与区域降雨量相同的估算方式得出一个空间平均值。

为解决雨量计测量净雨量时的不稳定性，可使用水槽或塑料薄膜。水槽可在更大的范围内有效收集净雨。塑料薄膜是大面积收集净雨的终极方案，但其也有缺陷。首先是安装和维护困难，特别是要确保塑料薄膜没有裂口；第二是地面上要有一个不透水层，以确保没有或很少的水进入土壤。这对短期研究来说不是问题，但对长期的研究来说，不透水层会使树木缺水而导致林冠发生改变。对于研究水量收支情况来说，这确实是个问题。

(3) 茎流量

测量茎流量的常见方法是在树干周围放置项圈，以捕捉所有从树干流出的水。在树皮光滑的树上，该方法相对操作简单。但在粗糙的树上（如在许多针叶树上）就非常困难。重要的是项圈要密封在树上，而且要足够大，大到可以容纳所有从树干流下来的水。套环也应向一侧倾斜，以便用翻斗式雨量计收集和测量。

3.3.2 测量降水量的替代方法

目前，直接估算面降水量是一个热点课题。有两种技术可以实现面降水量的估算：雷达和卫星遥感。这两种方法有许多相似之处，但它们在测量方向上有本质不同。雷达是通过从地球表面向上探测大气层来估计面降水量。卫星遥感是从太空向下探测地球表面来估算面降水量。

3.3.2.1 雷达

地面雷达的主要用途是天气预报，即跟踪雨云和锋面在地球表面的移动。雷达的几种功能都基于类似的原理。雷达是无线电探测和测距的缩写。一种电磁能量波从地面上的装置发出，记录波的反射量和返回时间。云中的水越多，则反射回地面并被雷达探测到的电磁能量就越大；反射波越快返回地面，则云离地面越近。这项技术的难点在于如何找到发射和探测的最佳电磁辐射波长。电磁波应被云中的液态水反射，而不是被大气中的气体反射。为了找到最适合的波长，学术界进行了大量的研究。研究表明，反射电磁波与降雨强度有强的相关性。因此，这可以作为估算降雨量的替代措施。如果需要准确估计降雨强度，则必须使用多个雨量计进行校准。但这导致一个问题：这项技术中，利用云的反射率并没有通用的公式可以得到降雨强度。每个站点需单独校准，这可能需要用到与云反射率一致的测量点数年的降雨量记录。同时，雷达的成本也限制了雷达的广泛使用，对于雨量计覆盖率差的地区，雷达并不是一个廉价的替代选项。

3.3.2.2　卫星遥感

卫星遥感与雷达有着本质的不同。这是因为卫星传感器探测的是云层的顶部而不是底部。经证实，如果云层顶部明亮且寒冷，则很有可能产生降雨。这些特性可以通过卫星传感器从太空观测到。

卫星传感器最常见的形式是无源感应（意思是它接收来自另一个来源的辐射，通常是太阳，而不是像雷达那样发出自身的辐射），并可探测可见光和红外波段的辐射。LANDSAT、SPOT 和 AVHRR 就是基于此类卫星平台，通过感应电磁光谱的可见光和红外部分，探测云的亮度（可见光）和温度（热红外）。"亮度温度"与降雨强度相关，可利用雨量计进行校准，类似于地面雷达。但这种方法的问题是：地面上的雪反射光和大气中的云反射光它们具有相似的亮度温度值，这会很难区分。

另一种可以利用的卫星传感器是无源微波。在太空中可以探测到地球发出的微波（低水平）。当地球表面和卫星传感器（即大气中的云）之间有液态水时，一些微波被水吸收。因此，如果卫星传感器没有检测到微波，就说明有云层（或地表其他水体）的存在。

尽管卫星遥感取得了一些突破，但是目前可用的最佳分辨率约为 10km×10km。因此，其在流域尺度的应用性很小。卫星遥感虽可以粗略地估计地区降水量，但离实际应用还有很长的路要走。研究表明，在雨量计密度较高的国家或地区，卫星遥感虽不能提高对区域降水量估计的准确度，但在雨量计覆盖率较低的地区，卫星遥感是一种有效的工具。几乎全世界都被卫星图像覆盖，这至少可以成为雨量计稀少的地区的一种选择。21 世纪初发射的新一代卫星平台可搭载多个传感器，可以同时测量可见光、红外线和微波波段。这大大提高了测量的精度，但这毕竟是一种间接测量降水量的方法，仍需要用雨量计对其进行校准。

3.4　降水的空间分布估计

在水文学中，我们通常要估计在一段时间内输送到给定区域（例如流域）的总水量，即这个地区的平均降水量：

$$P = \frac{1}{A} \iint_A p(x,y) \mathrm{d}x \mathrm{d}y \tag{3-1}$$

式中　P——流域的平均降水量；

　　　　A——流域面积；

$p(x, y)$——所考察时段的空间降水分布。

一旦 P 确定，就可以计算出输送到该地区的总水量为 PA。这里描述的降水量都是在小范围测量的降水量。事实上，人们需要了解一个大得多的地区（通常是一个流域）的降水量（见第 5 章）。因此，降水量需要从点测量转移到空间分布估计，这就需要采用某种形式的空间平均方法。空间平均应考虑雨量计分布的不均匀性和影响降雨分布的各种因素（如海拔、坡向和坡度）。

简单的算术平均值方法很少应用，因为这种方法仅适用于均匀分布且地形单一的流域中的雨量计采样样本。有多种统计方法可以解决空间分布问题。随着地理信息系统（GIS）的使用，计算变得相对容易。只有充分了解该技术的工作原理，才能理解各种技术的不足。这里将介绍 3 种方法：泰森多边形法、等高线法和等雨量线法。

3.4.1 泰森多边形法

泰森是 20 世纪初美国的一位工程师，他设计了一种简单的方法来克服流域内雨量计分布不均的问题。本质上，泰森多边形是在每一个代表性的区域均赋予一个雨量计。代表区域（多边形）的大小取决于每个雨量计与周围其他雨量计的距离。

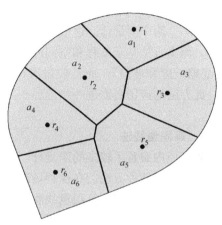

图 3-6　假想流域的泰森多边形

注：圆点为雨量计的位置。

每个多边形均在图上绘制；多边形的边界与每个雨量计等距，并且垂直（正交）于两个仪表之间的连线（图 3-6）。绘制完多边形后，即可得到雨量计周围多边形的面积。用式（3-2）计算空间平均降雨量（R）：

$$R = \sum_{i=1}^{n} \frac{r_i a_i}{A} \tag{3-2}$$

式中　r_i——i 号雨量计处的降雨量；
　　　a_i——雨量计 i 周围多边形的面积；
　　　A——流域总面积。

泰森多边形的面雨量值是各雨量计所测数值的加权平均值，权重基于每个代表区域（多边形）的大小。本方法仅适用于每个多边形内地形一致的情况，因此可假设降雨在每个多边形内的分布都是均匀的。这意味着考察流域降雨量之前就要确定按照这种方法对雨量计进行布点安装。

3.4.2 等高线法

众所周知，降雨量受海拔高度的影响（即海拔越高，降雨量越大）。因此，可以将流域海拔高度的分布用于估计降雨的空间分布。流域高程最简单的标记方法是等高线。图 3-7 为假想流域的海拔分布。等高线的数

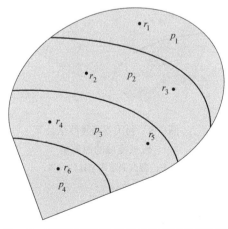

图 3-7　利用等高线法计算假想流域的面雨量

注：阴影区域位于两个轮廓线之间，降雨量是阴影区内两个仪表之间的平均值，圆点表示雨量计的位置。

值可从用平面仪测得的地图或地理信息系统（GIS）中的数字高程模型（DEM）地图中得到。

等高线法是根据两个高程之间的流域面积和实测降雨量计算加权平均值（式3-3）：

$$R = \sum_{j=1}^{m} r_j p_j \tag{3-3}$$

式中　r_j——两个等高线之间区域的平均降雨量；

　　　p_j——这些等高线之间的面积占整个流域面积的比例。

r_j 值为某一等高线间隔区域多个雨量计的平均值。如图 3-7 所示，阴影区（p_3）内有 r_4 和 r_5 两个雨量计。这里的 r_j 值就是 r_4 和 r_5 的平均值。泰森多边形法仅基于空间间距，而本方法基于高程和空间间距。等高线法假设海拔是影响降雨分布的唯一地形参数（即忽略坡度和坡向），而且假设海拔和降雨量之间的关系是线性的。然而，实际并非如此，此方法的应用尚需进一步论证。

3.4.3　等雨量线法

若流域内布置了较多的雨量计，平均降雨量的权重应直接由所测的降雨分布得到，而不是基于上所述两种方法。在这种情况下，可以通过在降雨量值之间进行插值法得到平滑的等降雨量曲线。传统的等雨量线法是先在地图上绘制，然后计算等雨量线之间的面积。空间平均值可由式（3-4）计算得到：

$$R = \frac{1}{A} \sum_{i=1}^{n} a_i r_i \tag{3-4}$$

式中　a_i——每个等雨量线之间的区域面积；

　　　r_i——等雨量线之间的平均降雨量。

本方法类似于等高线法。但曲线所连接的是相等的降雨量而不是高程（图 3-8）。

随着GIS的发展，等雨量线的插值计算和绘制都变得相对容易。插值法是将流域细分为小的网格单元，然后为每个单元格指定一个降雨值（可获得平滑的降雨曲面）。最简单的插值法是采用最近邻分析，是指网格区域的指定降雨量值与其最近的降雨量值成比例。克里格法更为复杂一些，它的每个单元格的插值是根据相邻雨量计之间的协方差的密切程度来推导的。

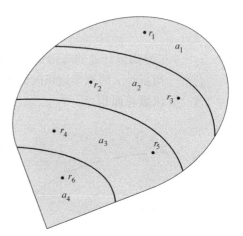

图 3-8　利用等雨量线法计算假想流域的面雨量

注：圆点表示雨量计的位置。

流域内特定点的可能降雨量也可以通过插值法从降雨面获得。在进行水文过程的数值模拟时，这个信息比流域的总降雨量更重要。

从降雨量的点测量到其空间分布平均值的估算均涉及典型的研究尺度问题。雨量计的测量范围（即雨量计表面积）远小于流域面积。将测量范围从点源扩大到整个流域是否可行？或者，是否应该有某种形式的换算系数来平衡其中的差异？这些问题将是水文学21世纪研究的重点问题。

Questions

3-1 Describe the different factors affecting the spatial distribution of precipitation at differing scales.

3-2 How are errors in the measurement of rainfall and snowfall minimized?

3-3 Compare and contrast different techniques for obtaining a spatially averaged precipitation value (including surrogate measures).

3-4 Why is scale such an important issue in the analysis of precipitation in hydrology?

3-5 Describe a field experiment (including equipment) required to measure the water balance beneath a forest canopy.

3-6 Discuss the role of spatial scale in assessing the importance of a forest canopy within a watershed.

3-7 Describe the different types of rainfall.

3-8 What are the problems associated with extrapolating rainfall gauge data to watersheds?

3-9 Precipitation is typically measured as a volume $[L^3]$ per unit area $[L^2]$, which has dimensions of length $[L]$. In the United States, the average annual precipitation varies from a minimum at Death Valley, California (1.6 inches), to a maximum on Mt. Waialeale on the island of Kauai in Hawaii (460 inches). What is the average annual precipitation (in millimeters, mm) at each of these locations? (You might want to review Appendix A on units, dimensions and conversions)

3-10 In the United States, stream discharge is often measured in units of cubic feet per second (ft^3/s, or "cfs"). In most other countries, discharge is measured in cubic meters per second (m^3/s). What is the equivalent flow (in m^3/s) of $18.2 ft^3/s$? (You might want to review Appendix A on units, dimensions and conversions)

3-11 In an average year, 1.0 meter of precipitation falls on a watershed with an area of 1000 (or 10^3) km^2.

(1) What is the volume of water received during an average year in cubic meters?

(2) What about in gallons?

Chapter 4 EVAPORATION

In terms of the water quantities transported on a global basis, evaporation is the second most important component of the hydrologic cycle, after precipitation.

4.1 Evaporation Mechanisms

As a physical phenomenon, evaporation is the transition of water from the liquid phase to the vapor phase. This transition requires first, an energy supply to provide water molecules the necessary kinetic energy to escape from the liquid surface; and second, some conditions to remove the escaped molecules from the immediate vicinity of the liquid surface thus preventing that they would return to condense.

Evaporation describes the net flow of water away from a surface. Water molecules also return to the evaporating surface by mass transport and diffusion processes. If the amount of vapor arriving equals the amount leaving, a steady state exists and no evaporation occurs. If there are more molecules arriving than leaving a surface, a net gain results, which is condensation. The vapor pressure of water molecules at the evaporating surface must exceed the vapor pressure in the atmosphere for evaporation to occur. Under natural circumstances, the vapor pressure of liquid water is mainly a function of its temperature although solute content, atmospheric pressure and water-surface curvature in capillaries can also be important. The vapor pressure of water molecules in the atmosphere is primarily a function of air temperature and humidity of the air. Figure 4-1 illustrates that a parcel of unsaturated air (A) must be cooled (A to C) or moisture must be added (A to B) before saturation occurs.

The vapor pressure gradient between evaporating surfaces and the atmosphere is the driving force that causes a net movement of water molecules. Such models assume that the vapor flowing away from evaporating surfaces is directly proportional to the vapor pressure deficit and inversely proportional to the resistance of air to the molecular diffusion and mass transport of water vapor. That is

$$E=\frac{e_s-e_a}{R_v} \tag{4-1}$$

Where E——the vapor flowed away from evaporating surfaces;

e_s——the vapor pressure of the evaporating surfaces;

e_a —— the vapor pressure of the atmosphere;

R_v —— the resistance of air to the molecular diffusion and mass transport of water vapor.

$e_s - e_a$ is the vapor pressure deficit between an evaporating surface and the atmosphere (see the difference between points A and B in Figure 4-1). Conceptual relationship of the evaporative process has been developed for complex surfaces such as plants and soils.

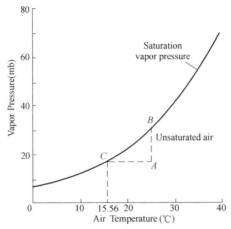

Figure 4-1 The Relationship of Saturation Vapor Pressure and Air Temperature
Source: BROOKS K N, FFOLLIOTT P F, MAGNER J A. Hydrology and the Management of watersheds (4th ed). 2013.

4.2 Evaporation from Water Surface

Evaporation from lakes, ponds, reservoirs, or swimming pools is determined only by energy and vapor flows. Because many areas of the world depend on reservoirs to provide municipal water supplies and water for irrigation, evaporation losses are important in determining whether the reservoir storage is sufficient to meet water demands. Although municipal water supply reservoirs are found in dry and wet regions alike, the greatest evaporation rates tend to occur in the driest regions where water is scarcer (Table 4-1).

Evaporation from Selected Lakes and Reservoirs in the United States and Australia　　Table 4-1

Location	Annual Evaporation (mm)
Lake Superior (1959~1966)	760
Pyramid Lake, Nevada	1118
Devils Lake, North Dakota	366
Oklahoma City (Reservoir)	1676
Seattle, Washington (Reservoir)	610
Salt Lake City, Utah (Reservoir)	1397
Lake Eyre, Australia	2134

Data Source: VAN DER LEEDEN F, TROISE F L, TODD D K. The Water Encyclo pedia (2nd ed). 1990.

Methods of determining lake evaporation include the water budget approach and the use of empirical relationships. Lake evaporation can be estimated with the water budget method if all components of the water budget except evaporation are either measured or estimated over a period of time (t):

$$E = Q_i + GW_i + P - Q_0 - GW_0 - S \tag{4-2}$$

where E——the lake evaporation (m^3/time);

P——the precipitation on the lake surface (m^3/time);

Q_i and Q_0——surface water flows into and out of the lake (m^3/time), respectively;

GW_i and GW_0——subsurface or groundwater flows into and out of the lake (m^3/time), respectively;

S——the change in storage (m^3/time).

Measuring all of the inflow and outflow components of a lake is difficult. Although it is straightforward to measure surface streams that enter and leave a lake, surface runoff directly into the lake from the surrounding area cannot be directly measured. Neither can the subsurface or groundwater flows be directly measured.

The most common method of estimating evaporation from a free-water body is by means of an evaporation pan. It is filled to a depth of 8 in. The water surface level is measured daily by a hook gage in a stilling well. The evaporation is computed as the difference in the observed levels adjusted for any precipitation during observation intervals. It has been observed that evaporation occurs more rapidly from a pan than from larger water bodies. A coefficient is accordingly applied to pan observations to derive the equivalent lake or reservoir evaporation, as shown in Equation (4-3).

$$E_L = KE_p \tag{4-3}$$

Where E_L——the evaporation from a water body;

E_p——the evaporation from the pan;

K——the pan coefficient.

Pan coefficients are normally computed on an annual basis, but monthly coefficients also have been used. The coefficient values range from 0.6 to 0.8 with an average of 0.7.

The standard evaporation pan in the United States, a National Weather Service Class A pan, is a metal cylinder with 122cm in diameter and 25cm deep. Water depth is maintained at 18~20cm and measured daily with a hook gauge in a stilling well (Figure 4-2). This pan sits above the surface (to lessen rain splash) and has either an instrument to record water depth or a continuous weighing device to measure changes in volume.

Figure 4-2   An Evaporation Pan

4.3   Evaporation from Soil Surface

Evaporation from a soil is a more complex phenomenon than evaporation from a water body. Given a bare, flat, wet soil surface, the water supply is initially unlimited and the amount of evaporation depends on the energy supply and the vapor pressure gradient as with a water body. When soils are exposed to the open atmosphere, sufficient vapor pressure gradients usually exist and are maintained, causing only the energy supply to limit evaporation from the wet and exposed soil. If a piece of transparent plastic sheeting was laid over this soil, evaporation would cease because vapor flow would be blocked even though energy and water flows would still converge at the active surface. In natural environments, however, energy inputs to the active surface increase the vapor pressure of water, steepening the vapor pressure gradient.

Evaporation proceeds at rates similar to those of free-water surfaces, assuming equal energy input in this case. As evaporation occurs, the lost water is replaced by water moving up from below the evaporating surface through the connecting water films around soil particles and through capillary pores. As the soil surface dries, the gradient in total water potential in the soil increases with water moving through the soil from the zone of higher potential (lower layer, wet zone) to the region of lower potential (upper layer, drier evaporating surface). Water films around the soil particles become thin, and the pathway through which water must move to reach the evaporating surface becomes more tortuous, thereby reducing hydraulic conductivity. After a period of drying, the rate of water flow through the soil limits the rate of evaporation at the soil surface.

Water flow in moist soil is primarily liquid but as soils dry, vapor diffusion through pores becomes more dominant. Water cannot move as rapidly through soils in vapor form as in liquid form. Therefore, deficient soil moisture limits evaporation at the active surface regardless of the energy input. With time, evaporation rates decrease because the water flow to the active evaporating surface is too slow to keep pace with the energy input. How much water will evaporate from a soil under these conditions depends largely on soil texture. Fine-textured soils retain pore water continuity at lower water contents than coarse-textured soils. Therefore, soil evaporation diminishes sooner in sandy soils than in clay soils, which have smaller pores that permit water films to remain intact for a longer period.

4.4 Transpiration

Transpiration is the most complex evaporative process. It requires a flow of liquid water to plant cell surfaces in leaves and exits through plant stomata in most plants. The rate of evaporation depends on the rate of vapor flow away from these surfaces in all instances. If one or more of these flows are changed, there is a corresponding change in the total evaporative loss from a surface.

Transpiration requires both energy and conditions that permit water vapor to flow away from evaporating or transpiring surfaces. Water molecules migrate from the liquid surface as a result of their kinetic energy. This transfer involves a change of state from liquid to vapor caused by energy inputs to the evaporating surface. Vapor flow is initially a diffusion process in which water molecules diffuse from a region of higher concentration (the evaporating surface or source) toward a region of lower concentration (a sink) in the atmosphere. Water molecules at the soil-atmosphere or leaf-atmosphere interface must firstly diffuse through the boundary layer. This is also the layer through which sensible heat is transferred by molecular conduction only. The boundary layer of air adjacent to evaporating and transpiring surfaces can be as thin as 1mm or less, but is at maximum thickness under still-air conditions. Wind and air turbulence reduce the boundary layer thickness but there is no turbulent flow in the boundary layer itself.

Once water molecules leave the boundary layer, they move into a turbulent zone of the atmosphere where further movement is primarily by mass transport, that is, turbulent eddy movement. In mass transport, whole parcels of air or eddies with water vapor and sensible heat flow in response to atmospheric pressure gradients, which cause the air parcels to flow both vertically and horizontally.

4.5 Evapotranspiration

Evaporation from water surface, soils, and plant interception and transpiration are considered collectively as evapotranspiration (ET). ET is the result of cumulative evaporation processes with each process requiring a change in the state of water from liquid to vapor and the net transfer of this vapor to the atmosphere. Therefore, there must be a flow of energy to the evaporating or transpiring surface before evaporation or transpiration can occur.

ET largely determines what proportion of precipitation input to a watershed becomes streamflow, and is influenced by land-use activities that alter vegetation and water bodies on the landscape. The water budget equation can be used to estimate ET over a period as follows:

$$ET = P - Q - S - l \tag{4-4}$$

where ET——the evapotranspiration (mm);

P——the precipitation (mm);

Q——the streamflow (mm);

S——the change for storage in the watershed, $S = S_2 - S_1$ (mm), where S_2 is the storage at the end of a period and S_1 is the storage at the beginning of a period;

l——the change in deep seepage, $l = l_o - l_i$ (mm), where l_o is the seepage out of the watershed and l_i is the seepage into the watershed.

The ratio ET/P is close to unity for dry climates, meaning that the ratio Q/P is small. ET/P is less in humid climates where the magnitude of ET is governed by available energy rather than the availability of water. Changes in vegetation that reduce ET will increase streamflow and/or groundwater recharge while increases in ET have the opposite effect.

Rates of ET influence water yield by affecting the antecedent water status of a watershed. High rates deplete water in the soil and in surface water impoundments, leaving more space that is available to store precipitation. Low ET rates leave less storage space in the soil and in surface water impoundments. The amount of storage space in a watershed affects the amount and, to some extent, the timing of streamflow resulting from precipitation events. ET is the result of cumulative evaporation processes with each process requiring a change in the state of water from liquid to vapor and the net transfer of this vapor to the atmosphere. Therefore, there must be a flow of energy to the evaporating or transpiring surface before evaporation or transpiration can occur.

Summary

Evapotranspiration from a water shed includes a variety of different vaporization processes, including evaporation from open water, soils and vegetation surfaces, and transpiration from plants. The evaporation process involves the transfer of water from a liquid state into a gaseous form in the atmosphere. Transpiration is the vaporization and transport of plant water from leaf chloroplasts to the atmosphere through the stomata, small cavities existing on leaf surfaces. It is difficult in practice to separate evaporation (from wet surfaces) and transpiration (water evaporating inside plants) from each other. Fundamentally, it is through the process of evapotranspiration that the sun's energy is introduced to drive the hydrological cycle watershed.

Translation of Some Sections
部分章节参考译文

4.1 蒸发机理

蒸发是水从液相转变为气相的一种物理现象。首先，蒸发需要能量供应，为水分子从液体表面逸出提供所需的动能；其次，蒸发需要某种条件，将逸出的分子从液体表面附近移除，从而防止它们返回冷凝状态。

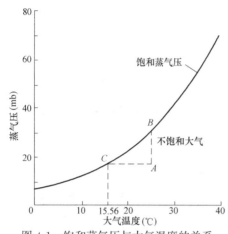

图 4-1 饱和蒸气压与大气温度的关系
来源：BROOKS K N, FFOLLIOTT P F, MAGNER J A. Hydrology and the Management of Watersheds (4th ed). 2013.

蒸发是水从表面流出的净流量。水分子会在质量的迁移和扩散的同时返回蒸发表面。如果返回的蒸汽量等于离开的蒸汽量，则此时是一个稳定状态，不会发生蒸发现象。如果返回的分子比离开表面的分子多，就会产生一个净增量，这就是冷凝。当液态水表面的蒸气压超过大气中的蒸气压，蒸发才能发生。在自然条件下，液态水的溶质含量、大气压力和毛细管水面曲率会影响其蒸气压，但蒸气压主要受温度控制。大气中水分子的蒸气压主要是大气温度和湿度的函数。图 4-1 说明了不饱和大气（A）达到饱和有两条途径：一是冷却（A 到 C）；二是增加大气湿度（A 到 B）。

蒸发表面和大气之间的蒸气压梯度是导致水分子净运动的驱动力。假设蒸发量与蒸发表面的蒸气压与大气蒸气压的压差成正比，与大气对水蒸气分子扩散和质量传输的阻力成反比。因此，蒸发量为：

$$E=\frac{e_s-e_a}{R_v} \tag{4-1}$$

式中 E——蒸发量；
e_s——蒸发表面的蒸气压；
e_a——大气的蒸气压；
R_v——大气对水蒸气分子扩散和质量传输的阻力。

e_s-e_a 是蒸发表面和大气之间的蒸气压差（见图 4-1 中 A 点和 B 点之间的差值）。蒸发过程已从简单的水面，发展为复杂的表面（如植物和土壤）。

4.4 蒸腾作用

蒸腾作用是最复杂的蒸发过程。对大多数植物来说，它是指一股液态水流向植物叶片中的细胞表面，并通过叶片上的气孔排出。蒸发速率取决于离开这些表面的蒸汽流速率。

若这些蒸汽流中的一个或多个环节发生变化,则总蒸发损失也会相应变化。

蒸腾作用的发生需要满足一定的能量和允许水蒸气从蒸发或蒸腾表面流出的条件。水分子具有从液体表面迁移的动能,而能量输入包括在蒸发或蒸腾表面引起水的状态从液体到气体变化时所需的能量。蒸汽流最初是一个扩散的过程,在此过程中,水分子从较高浓度的区域向大气中较低浓度的区域扩散。土壤-大气或叶-大气界面的水分子首先通过边界层来传递焓。与蒸发和蒸腾表面相邻的空气边界层可以薄到1mm左右,甚至更薄。边界层在静止空气下的厚度最大。风和空气湍流可降低边界层厚度,但边界层内没有湍流流动。

一旦水分子离开边界层,它们就会进入大气湍流区。接下来水分子主要是进行质量输运,即做湍流运动。质量输运是指空气分子和焓在大气压力梯度的作用下所做的垂直和水平方向上的流动。

4.5 蒸 散 发

水面蒸发、土壤蒸发和植物蒸腾统称为蒸散发。蒸散发是水的状态从液体变为水蒸气,水蒸气到大气的净转移的一个累积过程。因此,在蒸发或蒸腾发生之前,必须有能量流向蒸发或蒸腾表面。

蒸散发在很大程度上决定了流域内降水的产流比,也会受到如植被、景观水体的改变等土地利用活动的影响。水量平衡方程可用于估算一段时间内流域的蒸散发量:

$$ET = P - Q - S - l \tag{4-4}$$

式中 ET——流域蒸散发量(mm);

P——流域降水量(mm);

Q——流域的径流量(mm);

S——流域蓄水量的变化,$S = S_2 - S_1$ (mm),其中 S_2 为末期蓄水量,S_1 为初期蓄水量;

l——流域深层渗水量的变化,$l_o - l_i$ (mm),其中 l_o 是渗出量,l_i 是渗入量。

干燥气候有利于蒸发,此时的 ET/P 接近于1,Q/P 很小。潮湿气候不利于蒸发,ET/P 较低,此时 ET 是由能量而不是湿度决定。若 ET 降低,有利于河流补给地下水;若 ET 增加,则相反。

蒸散发对产流量的影响是通过影响流域的先前水分状况来实现的。若土壤和地表水体中的水分高速消耗,则会留出更多的储水空间;若低速消耗,则会留出较少的储水空间。流域蓄水空间的大小也在一定程度上影响着径流产生所需的时间。

Questions

4-1 Give a detailed account of the factors influencing evaporation rate above a forest canopy.

4-2 Outline the major evaporation estimation techniques and compare their effectiveness for your local environment.

4-3 Describe the factors that restrict actual evaporation (evapotranspiration) from equaling potential evaporation in a humid-temperate climate.

4-4 What does evaporation of water from water surfaces depend on?

Chapter 5 RUNOFF

The amount of water within a river or stream is of great interest to hydrologists. It represents the end product of all the other processes in the hydrological cycle and is where the largest amount of effort has gone into analysis of historical records. Runoff is a loose term that covers the movement of water to a channelized stream, after it has reached the ground as precipitation. The movement can occur either on or below the surface and at differing velocities. Once the water reaches a stream, it moves towards the oceans in a channelized form, of which the process referred to as streamflow or riverflow. Streamflow is expressed as discharge: the volume of water over a defined time period. The SI units for discharge are m^3/s (cumecs). A continuous record of streamflow is called a hydrograph(Figure 5-1). Although we think of this as continuous measurement, it is normally either an averaged flow over a time period or a series of samples (e. g. hourly records, daily records).

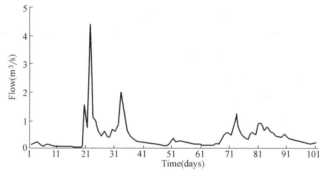

Figure 5-1 A Typical Hydrograph, Taken from the River Wye,
Wales for a 100-day Period during the Autumn of 1995
Source: FITTS C R. Groundwater Science. 2002.

In Figure 5-1, there are a series of peaks between periods of steady, much lower flows. The hydrograph peaks are referred to as peakflow, stormflow or even quickflow. They are the water in the stream during and immediately after a significant rainfall event. The steady periods between peaks are referred to as baseflow or sometimes slowflow. The shape of a hydrograph, and in particular the shape of the stormflow peak, is influenced by the storm characteristics (e. g. rainfall intensity and duration) and many physical characteristics of the upstream watershed. In terms of watershed characteristics, the largest influence is exerted by watershed size, but other factors include slope angles, shape of watershed, soil type, vegetation type and percentage cover, degree of urbanization and the antecedent soil moisture.

Figure 5-2 shows the shape of a storm hydrograph in detail. Several important hydrological

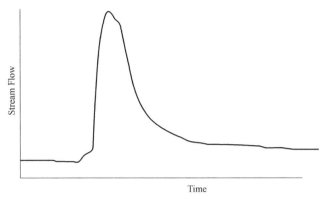

Figure 5-2 Demonstration Storm Hydrograph
Source: FITTS C R. Groundwater Science. 2002.

terms can be seen in this diagram. The rising limb of the hydrograph is the initial steep part leading up to the highest or peakflow value. The water contributing to this part of the hydrograph is from channel precipitation (i. e. rain that falls directly onto the channel) and rapid runoff mechanisms. Some texts claim that channel precipitation shows up as a preliminary blip before the main rising limb. In reality this is very rarely observed, a factor of the complicated nature of storm runoff processes. The recession limb of the hydrograph is after the peak and is characterized by a long, slow decrease in streamflow until the baseflow is reached again. The recession limb is attenuated by two factors: storm water arriving at the mouth of a watershed from the furthest parts, and the arrival of water that has moved as underground flow at a slower rate than the streamflow.

Exactly how water moves from precipitation reaching the ground surface to channelized streamflow is one of the most intriguing hydrological questions, and one that cannot be answered easily (Figure 5-3).

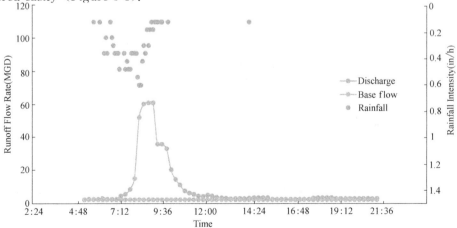

Figure 5-3 A Rainfall-discharge Chart
Note: MGD means million gallons per day.
Source: LI S, KAZEMI H, ROCKAWAY T D. Performance Assessment of Stormwater GI Practices: Using Artificial Neural Networks. 2019.

Much research effort in the past hundred years has gone into understanding runoff mechanisms; considerable advances have been made, but there are still many unanswered questions. The following section describes how it is believed that runoff occurs, but there are many different scales at which these mechanisms are evident and they do not occur everywhere.

5.1 Runoff Mechanism

Overland flow (Q_o) is the water which runs across the surface of the land before reaching the stream. In the subsurface, throughflow (Q_t) (some authors refer to this as lateral flow) occurs in the shallow subsurface, predominantly, although not always, in the unsaturated zone. Groundwater flow (Q_g) is in the deeper saturated zone. All of these are runoff mechanisms that contribute to streamflow. The relative importance of each is dependent on the watershed under study and the rainfall characteristics during a storm.

5.1.1 Overland Flow

Some of the earliest research work on how overland flow occurs was undertaken by Robert Horton (1875~1945). He hypothesized that overland flow occurred when the rainfall rate was higher than the infiltration rate of a soil. Horton went on to suggest that under these circumstances the excess rainfall collected on the surface before travelling towards the stream as a thin sheet of water moving across the surface. Under this hypothesis, it is the infiltration rate of a soil that acts as a controlling barrier or partitioning device. Where the infiltration capacity of a soil is low, overland flow occurs readily. This type of overland flow is referred to as infiltration excess overland flow or Hortonian overland flow (see Chapter 7).

In addition to the infiltration capacity information, it is extremely rare to find a thin sheet of water moving over the surface during a storm event. It was observations such as those by Hursh and others that led to a general revision of Horton's hypothesis. Betson proposed the idea that within a watershed there are only limited areas that contribute overland flow to a storm hydrograph.

This is referred to as the partial areas concept. Betson did not challenge the role of infiltration excess overland flow as the primary source of stormflow, but did challenge the idea of overland flow occurring as a thin sheet of water throughout a watershed. Hewlett and Hibbert were the first to suggest that there might be another mechanism of overland flow occurring. This was particularly based on the observations from the eastern USA that during a storm it was common to find all the rainfall infiltrating a soil. Hewlett and Hibbert hypothesized that during a rainfall event all the water infiltrated the surface.

So who was right: Horton, Betson or Hewlett and Hibbert? The answer is that both were. Table 5-1 provides a summary of the ideas for storm runoff generation described here. It is now accepted that saturated overland flow (Hewlett and Hibbert) is the dominant overland flow mechanism in humid, midlatitude areas. It is also accepted that the variable source areas concept is the most valid description of stormflow processes. However, where the infiltration capacity of a soil is low or the rainfall rates are high, Hortonian overland flow does occur. In Table 5-1, it can be seen that there are times when rainfall intensities will exceed infiltration rates under natural circumstances. In arid and semi-arid zones it is not uncommon to find extremely high rainfall rates that can lead to infiltration excess overland flow and rapid flood events; this is called flash flooding.

A Summary of the Ideas on How Stormflow is Generated in a Watershed   Table 5-1

	Horton	Betson	Hewlett and Hibbert
Infiltration	Controls overland flow	Controls overland flow	All rainfall infiltrates
Overland Flow Mechanism	Infiltration excess	Infiltration excess	Saturated overland flow
Contributing Area	Uniform throughout the watershed	Restricted to certain areas of the watershed	Contributing area is variable in time and space

Examples of low infiltration rates can be found with compacted soils (e. g. from vehicle movements in an agricultural field), on roads and paved areas, on heavily crusted soils and what are referred to as hydrophobic soils.

5.1.2 Subsurface Flow

Under the variable source areas concept there are places within a watershed that contribute overland flow to the storm hydrograph. When we total up the amount of water found in a storm hydrograph, it is difficult to believe that it has all come from overland flow, especially when this is confined to a relatively small part of the watershed (i. e. variable source areas concept). The manner in which the recession limb of a hydrograph attenuates the stormflow suggests that it may be derived from a slower movement of water: subsurface flow. In addition to this, tracer studies looking at where the water has been before entering the stream as stormflow have found that a large amount of the storm hydrograph consists of "old water". This old water has been sitting in the soil, or as fully saturated groundwater, for a considerable length of time and yet enters the stream during a storm event. There have been several theories put forward to try to explain these findings, almost all involving throughflow and groundwater.

Throughflow is a general term used to describe the movement of water through the unsaturated zone; normally this is the soil matrix. Once water infiltrates the soil surface it continues to move, either through the soil matrix or along preferential flow paths (referred to as lateral or preferential flow). The rate of soil water movement through a saturated soil matrix is described by Darcy's law (see Chapter 7) and the Richard's approximation of Darcy's law when below saturation. Under normal, vertical, infiltration conditions the hydraulic gradient has a value of -1 and the saturated hydraulic conductivity is the infiltration capacity. Once the soil is saturated, the movement of water is not only vertical. With a sloping water table on a hillslope, water moves down slope. However, the movement of water through a saturated soil matrix is not rapid. In order for throughflow to contribute to storm runoff, there must be another mechanism (other than matrix flow) operating.

One of the first theories put forward concerning the contribution of throughflow to a storm hydrograph was by Horton and Hawkins (Horton here was a different person from the proposer of Hortonian overland flow). They proposed the mechanism of translatory or piston flow to explain the rapid movement of water from the subsurface to the stream. They suggested that as water enters the top of a soil column it displaces the water at the bottom of the column (i.e. old water), and the displaced water enters the stream. The analogy is drawn to a piston where pressure at the top of the piston chamber leads to a release of pressure at the bottom. The release of water to the stream can be modelled as a pressure wave rather than tracking individual particles of water. Piston flow has been observed in laboratory experiments with soil columns.

At first glance, the simple piston analogy seems unlikely in a real-life situation since impermeable sides do not bound a hillslope in the same way as a piston chamber. However, the theory is not as farfetched as it may seem, as the addition of rainfall infiltrating across a complete hillslope is analogous to pressure being applied from above and in this case, the boundaries are upslope (i.e. gravity) and the bedrock below. Brammer and McDonnell suggest that this may be a mechanism for the rapid movement of water along the bedrock and soil interface on the steep watershed of Maimai in New Zealand. In this case, it is the hydraulic gradient created by an addition of water to the bottom of the soil column, already close to saturated, that forces water along the base where hydraulic conductivities are higher.

Ward draws the analogy of a thatched roof to describe the contribution of subsurface flow to a stream. When straw is placed on a sloping roof, it is very efficient at moving water to the bottom of the roof (the guttering being analogous to a stream) without visible overland flow. This is due to the preferential flow direction along, rather than between, sloping straws. Measurements of hillslope soil properties do show a higher hydraulic conductivity in the downslope rather than vertical direction. This would account for a movement of water downslope as throughflow, but it is still bound up in the soil matrix and reasonably slow.

There is considerable debate on the role of macropores in the rapid movement of water through the soil matrix. Macropores are larger pores within a soil matrix, typically with a diameter greater than 3mm. They may be caused by soils cracking, worms burrowing or other biotic activities. The main interest in them from a hydrologic point of view is that they provide a rapid conduit for the movement of water through a soil. The main area of contention concerning macropores is whether they form continuous networks allowing rapid movement of water down a slope or not. There have been studies suggesting macropores as a major mechanism contributing water to stormflow, but it is difficult to detect whether these are from small areas on a hillslope or continuous throughout.

The role of macropores in runoff generation is still unclear. Although they are capable of allowing rapid movement of water towards a stream channel, there is little evidence of networks of macropores moving large quantities of water in a continuous fashion. Where macropores are known to have a significant role is in the rapid movement of water to the saturated layer, which may in turn lead to piston flow.

5.1.3 Baseflow

In sharp contrast to the storm runoff debate, there is consensus that the major source of baseflow is groundwater, and to a lesser extent throughflow. This water has infiltrated the soil surface and moved towards the saturated zone. Once in the saturated zone, it moves downslope, often towards a stream. A stream or lake is often thought to occur where the regional water table intersects the surface, although this may not always be the case. In Chapter 7, the relationship between groundwater and streamflow has been explained (see Figure 7-10). However, in general, baseflow is provided by the slow seepage of water from groundwater into streams. This will not necessarily be visible (e.g. springs) but can occur over a length of streambank and bed, and is only detectable through repeated measurement of streamflow down a reach.

5.1.4 Channel Flow

Once water reaches the stream it will flow through a channel network to the main river (Figure 5-4). The controls over the rate of flow of water in a channel are to do with the volume of water present, the gradient of the channel, and the resistance to flow experienced at the channel bed. This relationship is described in uniform flow formulae such as the Chezy and Manning Equations. The resistance to flow is governed by the character of the bed surface. Boulders and vegetation will create a large amount of friction, slowing the water down as it passes over the bed.

In many areas of the world, the channel network is highly variable in time and space. Small channels may be ephemeral and in arid regions will frequently only flow dur-

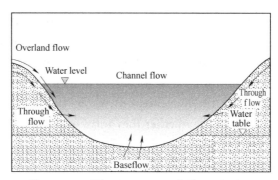

Figure 5-4 The Generation of Channel Flow

Source: DAVIS M L, MASTEN S J. Principles of Environmental Engineering and Science (3rd ed). 2014.

ing flood events. The resistance to flow under these circumstances is complicated by the infiltration that will be occurring at the waterfront and bed surface. The first flush of water will infiltrate at a much higher rate as it fills the available pore space in the soil/rock at the bed surface. This will remove water from the stream and slow the waterfront down as it creates a greater friction surface. Under a continual flow regime, the infiltration from the stream to ground will depend on the hydraulic gradient and the infiltration capacity.

5.2 Velocity Distribution in a Stream Section

Streamflow is the primary mechanism by which water moves from upland watersheds to ocean basins. Streamflow is the principal source of water for municipal water supplies, irrigated agricultural production, industrial operations and recreation. Sediments, nutrients, heavy metals, and other pollutants are also transported downstream in streamflow. The velocity in a stream section is not uniformly distributed, due to the presence of the free surface and friction along the stream wall. It varies both across the width and along the depth. Figure 5-5 indicates the general pattern of velocity distribution in a stream channel.

The maximum velocity usually occurs below the free surface near the center of the channel section. The velocity decreases toward the banks (Figure 5-5). Also, the closer to the banks, the deeper the point of highest velocity in a vertical section is. Factors that affect the velocity distribution are the shape of the section, the roughness of the channel, and

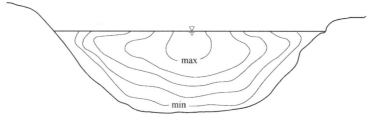

Figure 5-5 Typical Velocity Distribution in a Stream Section

Source: GUPTA R S. Hydrology and Hydraulic Systems (4th ed). 2017.

the presence of bends. The surface wind has very little effect. A spiral type of motion has been observed in laboratory investigations. In natural rivers, the spiral motion is usually so weak that its effect is practically eliminated by the channel friction.

5.3 Measuring Streamflow

Streamflow or discharge can be considered as a water velocity passing through a cross-sectional area. Flooding occurs when streamflow discharge exceeds the capacity of the channel. Therefore, streamflow measurements or methods for predicting streamflow characteristics are needed for various purposes. Streamflow measurement can be subdivided into two important subsections: instantaneous and continuous techniques.

5.3.1 Instantaneous Streamflow Measurement

The velocity-area method measures the stream velocity, the stream cross-sectional area and multiplies the two together:

$$Q = vA \tag{5-1}$$

where Q——the streamflow or discharge (m^3/s);
v——the average velocity (m/s) of water passing through a cross-sectional area;
A——the cross-sectional area (m^2).

In practice, this is carried out by dividing the stream into small sections and measuring the velocity of flow going through each cross-sectional area and applying Equation (5-2).

$$Q = \sum_{i=1}^{n} v_i a_i \tag{5-2}$$

where Q——the streamflow or discharge (m^3/s);
v_i——the velocity (m/s) measured in the cross-sectional area of trapezoidal i;
a_i——the area (m^2) of the trapezoid i (usually estimated as the average of two depths multiplied by the width between);
n——the number of sections.

The number of cross-sectional areas that is used in a discharge measurement depends upon the width and smoothness of streambed. If the bed is particularly rough, it is necessary to use more cross-sectional areas so that the estimates are as close to reality as possible (the discrepancy is noted between the dashed and solid lines in Figure 5-6).

The water velocity measurement is usually taken with a flow meter (Figure 5-7), which is a form of propeller inserted into the stream and records the number of propeller turns with time. In the velocity-area method, it is necessary to assume that the velocity measurement is representative of all the velocities throughout the cross-sectional area. It is not

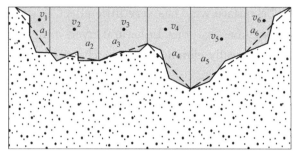

Figure 5-6　The Velocity-area Method of Streamflow Measurement

Note: The black circles indicate the position of velocity readings. Dashed lines represent the triangular or trapezoidal cross-sectional area through which the velocity is measured.

Source: DAVIE T. Fundamentals of Hydrology (2nd ed). 2008.

Figure 5-7　Flow Meter

normally possible to take multiple measurements so an allowance has to be made for the fact that the water travels faster along the surface than nearer the streambed. This difference in velocity is due to friction exerted on the water as it passes over the streambed, slowing it down. As a general rule of thumb, the sampling depth should be 60% of the stream depth——that is, in a stream that is 1m deep, the sampling point should be 0.6m below the surface or 0.4 m above the bed (Figure 5-8). In a deep river it is good practice to take two measurements (one at 20% and the other at 80% of depth) and average the two.

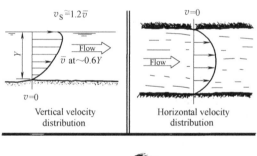

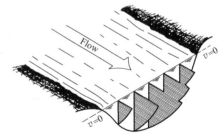

Figure 5-8　Measurements of Stream Channel Cross Sections and
Velocities Needed to Estimate the Mean Velocity of a Stream

Source: BROOKS K N, FFOLLIOTT P F, MAGNER J A. Hydrology and the Management of Watersheds (4th ed). 2013.

Where there is no velocity meter available, it may be possible to make a very rough estimate of stream velocity using a float in the stream (i.e. the time it takes to cover a measured distance). When using this method, allowance must be made for the fact that the float is travelling on the surface of the stream at a faster rate than water closer to the streambed.

The velocity-area method is an effective technique for measuring streamflow in small rivers, but its reliability is heavily dependent on the sampling strategy. The technique is also less reliable in small, turbid streams with a rough bed (e.g. mountain streams).

5.3.2 Continuous Streamflow Measurement

The methods of instantaneous streamflow measurement described above only allow a single measurement to be taken at a location. Although this can be repeated at a future date, it requires a continuous measurement technique to give the data for a hydrograph. There are four different techniques that can be used for this method: stage discharge relationship, flumes and weirs, radar stage measurement and ultrasonic flow gauging.

(1) Stage Discharge Relationship

River stage is another term for the water level or height. Where multiple discharge measurements have been taken (i.e. repeat measurements using velocity-area method), it is possible to draw a relationship between river stage and discharge: the so-called rating curve. An example of a rating curve is shown in Figure 5-9. This has the advantage of allowing continuous measurement of river stage (a relatively simple task) that can then be equated to the actual discharge. The stage discharge relationship is derived through a series of velocity-area measurements at a particular site while at the same time recording the stage with a stilling well (Figure 5-10). In Figure 5-10, the height of water is measured in the well immediately adjacent to the river. As can be seen in Figure 5-9, the rating curve is non-linear, a reflection of the riverbank profile. As the river fills up between banks, it takes a greater volume of water to cause a change in stage than at low levels.

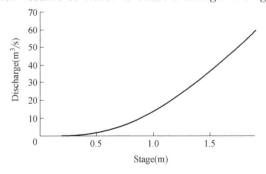

Figure 5-9 A Rating Curve for the River North Esk in Scotland based on
Stage (Height) and Discharge Measurements from 1963~1990
Source: DAVIE T. Fundamentals of Hydrology (2nd ed). 2008.

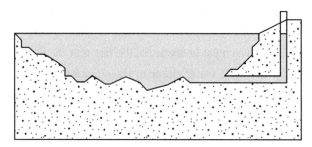

Figure 5-10 Stilling Well to Provide a Continuous Measurement of River Stage (Height)
Note: The height of water is measured in the well immediately adjacent to the river.
Source: DAVIE T. Fundamentals of Hydrology (2nd ed). 2008.

An accurate stage discharge relationship is dependent on frequent and accurate measurement of river discharge, and a static riverbed profile. If the riverbed profile changes (e. g. during a large flood event it may get scoured out or new sedimentis deposited), the stage discharge relationship will change and the historic relationship will no longer be valid. This assumption of a static riverbed profile can sometimes be problematic, leading to the installation of a concrete structure (e. g. flume or weir) to maintain stability.

One of the difficulties with the stage discharge relationship is that the requirement of frequent measurements of river discharge lead to many measurements taken during periods of low and medium flow but very few during flood events. This is for the double reason that: floods are infrequent and unlikely to be measured under a regular monitoring programme; and the danger of streamflow gauging during a flood event exists. The lack of data at the extreme end of the stage-discharge curve may lead to difficulties in interpreting data during peak flows.

The error involved in estimating peak discharge from a measured stage discharge relationship will be much higher at the high flow end of curve. When interpreting data derived from the stage discharge relationship, the hydrologist should bear in mind that it is the stream stage that is being measured and from this stream discharge is inferred (i. e. it is not a direct measurement of stream discharge).

(2) Flumes and Weirs

Flumes and weirs utilize the stage discharge relationship described above but go a step further towards providing a continuous record of river discharge. If we think of stream discharge as consisting of a river velocity flowing through a cross-sectional area (as in the velocity profile method), then it is possible to isolate both of these terms separately. This is what flumes and weirs, or stream gauging structures, attempt to do (Figure 5-11).

The first part to isolate is the stream velocity. The way to do this is to slow a stream down

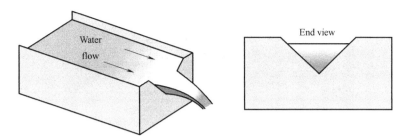

Figure 5-11　Weir for Stage Measurement
Source: DAVIS M L, MASTEN S J. Principles of Environmental Engineering and Science (3rd ed). 2014.

(or, in some rare cases, speed a stream up) so that it flows with constant velocity through a known cross-sectional area. The critical point is that in designing a flume or weir, the river flows at the same velocity (or at least a known velocity) through the gauging structure irrespective of how high the river level is. Although this seems counter-intuitive (rivers normally flow faster during flood events), it is achievable if there is an area prior to the gauging structure that slows the river down: a stilling pond.

The second part of using a gauging structure is to isolate a cross-sectional area. To achieve this, a rigid structure is imposed upon the stream so that it always flows through a known cross-sectional area. In this way, a simple measure of stream height through the gauging structure will give the cross-sectional area. Stream height is normally derived through a stilling well, as described in Figure 5-10, except in this case there is a regular cross-sectional area.

Once the velocity and cross-sectional area are kept fixed, the rating curve can be derived through a mixture of experiment and hydraulic theory.

(3) Radar Stage Measurement
The radar sensor is a self-contained unit having a hornlike transmitting device as shown in Figure 5-12. It operates by transmitting electromagnetic waves through the air. The waves reflect off the water surface and echo back to the sensor. The system calculates the distance on the basis of the time interval between the transmission and the echo. The sensor is internally programmed to convert the distance to water surface to the units of stage. The stage, as sensed by the radar, is transmitted to the data-collection platform by a hard-wired connection for storage and transmittal. The sensor is mounted above the water surface at a bridge or other suitable site. It has a solid, relatively light, water-proof housing that is easy to install and maintain. The radar sensor does not occupy space within a stream section and is not impacted by channel changes and debris. It can get over the flood stage. However, the device is not applicable to an iced river and needs to have a structure over the river for installation.

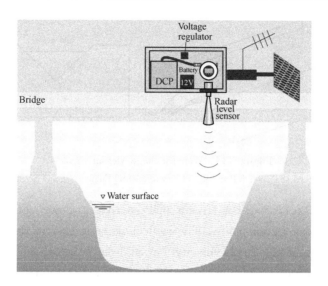

Figure 5-12 Schematic of a Radar-level Sensor used to Measure Stage
Source: TURNIPSEED D P, SAUER V B. Discharge
Measurements at Gaging Stations. 2010.

(4) Ultrasonic Flow Gauging

Recent technological developments have led to the introduction of a method of measuring stream discharge using the properties of sound wave propagation in water. The method actually measures water velocity, but where the streambed cross-sectional area is known (and constant) the instrumentation can be left in place and combined with measurements of stage to provide a continuous measurement of river discharge. There are two types of ultrasonic flow gauges that work in slightly different ways.

The first method measures the time taken for an ultrasonic wave emitted from a transmitter to reach a receiver on the other side of a river. The faster the water speed is, the greater the deflection of the wave path will be and the longer it will take to cross the river. Sound travels at approximately 1500m/s in water (dependent on water purity and depth) so the instrumentation used in this type of flow gauging needs to be extremely precise and be able to measure in nanoseconds. This type of flow gauging can be installed as a permanent device but needs a width of river greater than 5m and becomes unreliable with a high level of suspended solids.

The second method utilizes the Doppler Effect to measure the speed of particles being carried by the stream. At an extremely simple level, this is a measurement of the wavelength of ultrasonic waves that bounce off suspended particles——the faster the particle is, the shorter the wavelength will be. This type of instrument works in small streams (less than 5m width) and requires some suspended matter.

5.4 Floods

5.4.1 Introduction

Floods are a frequently occurring event around the world. The term flood is difficult to define except in the most general of terms. In a river, a flood is normally considered as an inundation of land adjacent to a river caused by a period of abnormally large discharge or encroachment by the sea, but even this definition is fraught with inaccuracy. Flooding may occur from sources other than rivers (e. g. the sea and lakes), and "abnormal" is difficult to pin down, particularly within a timeframe. Floods come to our attention through the amount of damage that they cause, and for this reason, they are often rated on a cost basis rather than on hydrological criteria.

Hydrological and monetary assessments of flooding often differ markedly because the economic valuation is highly dependent on location. If the area of land inundated by a flooding river is in an expensive region with large infrastructure, then the cost will be considerably higher than, say, for agricultural land. Two examples of large-scale floods during the 1990s illustrate this point. In 1998, floods in China caused an estimated RMB ¥140 billion of damage with over 15 million people being displaced and 3000 lives lost. This flood was on a similar scale to one that occurred in the same region during 1954. A much larger flood (in a hydrological sense) in the Mississippi and Missouri rivers during 1993 resulted in a similar economic valuation of loss (US\$15~20 billion) with 48 lives lost. The flood was the highest in the hydrological record and had an average recurrence interval of between 100 and 500 years.

5.4.2 Influences on Flood Size

The extent and size of the flood can often be related to other contributing factors that increase the effect of high rainfall. Some of these factors are described here but all relate back to concepts introduced in earlier chapters detailing the processes found within the hydrological cycle. Many of the factors discussed here have an influence at the small scale (e. g. hillslopes or small research watersheds of less than 10km^2) but not at the larger overall watershed scale.

(1) Antecedent Soil Moisture

The largest influence on the size of a flood, apart from the amount and intensity of rainfall, is the wetness of the soil immediately prior to the rainfall or snow melt occurring. As described on section 7.5.1, the amount of infiltration into a soil and subsequent storm runoff are highly dependent on the degree of saturation in the soil. Almost all major flood events are heavily influenced by the amount of rainfall that has occurred prior to the actual flood-causing rainfall.

(2) Deforestation

The effect of trees on runoff has already been described, particularly with respect to water resources. There is also considerable evidence that a large vegetation cover, such as forest, decreases the severity of flooding. There are several reasons for this. The first has already been described, in that trees provide an intercepting layer for rainfall and therefore slow down the rate at which the water reaches the surface. This will lessen the amount of rainfall available for soil moisture, and therefore, the antecedent soil moisture may be lower under forest than for an adjoining pasture. The second factor is that forests often have a high organic matter in the upper soil layers that, as any gardener will tell you, is able to absorb more water. Again, this lessens the amount of overland flow, although it may increase the amount of throughflow. Finally, the infiltration rates under forest soils are often higher, leading again to less saturation excess overland flow. The removal of forests from a watershed area will increase the propensity for a river to flood and increase the severity of a flood event. Conversely, the planting of forests on a watershed area will decrease the frequency and magnitude of flood events.

(3) Urbanization

Urban areas have a greater extent of impervious surfaces than in most natural landforms. Consequently, the amount of infiltration excess overland flow is high. In addition to this, urban areas are often designed to have a rapid drainage system, taking the overland flow away from its source. This network of gutters and drains frequently leads directly to a river drainage system, delivering more floodwater in a faster time. Where extensive urbanization of a watershed occurs, flood frequency and magnitude increase. studies show a massive increase in flood magnitude for an urban watershed when compared to a similar rural watershed. Urbanization is another influence on flooding that is most noticeable at the small scale. This is mostly because that the actual percentage area covered by impermeable urban areas in a larger river watershed is still very small in relation to the amount of permeable non-modified surfaces.

(4) River Channel Alterations

Geomorphologists traditionally view a natural river channel as being in equilibrium with the river flowing within it. This does not mean that a natural river channel never floods, but rather that the channel has adjusted in shape in response to the normal discharge expected to flow through it. When the river channel is altered in some way, it can have a detrimental effect on the flood characteristics for the river. In particular, channelization using rigid structures can increase flood risk. Ironically, channelization is often carried out to lessen flood risk in a particular area. This is frequently achieved, but in doing so water is passed on downstream at a faster rate than normal, increasing the flood risk further downstream. If there is a natural floodplain further downstream, this may not be a problem. But if there is not, downstream riparian zones will be at greater risk.

(5) Land Drainage

It is common practice in many regions of the world to increase agricultural production through the drainage of "swamp" areas. The drainage of these regions provides for rapid removal of any surplus water (i. e. not needed by plants). Drained land will be drier than that it might be expected naturally and therefore, less storm runoff might be assumed. This is true in small rainfall events but the rapid removal of water through subsurface and surface drainage leads to flood peaks in the river drainage system where normally the water would have been slower to leave the land surface. Therefore, although the drainage of land leads to an overall drying out of the affected area, it can also lead to increased flooding through rapid drainage.

(6) Climate Change

In recent years, any flooding event has led to a clamor of calls to explain the event in terms of climatic change. This is not easy to do as climate is naturally so variable. What can be said though is that river channels slowly adjust to changes in flow regime, which may in turn be influenced by changes in climate. Many studies have suggested that future climate change will involve greater extremes of weather, including more high intensity rainfall events. This is likely to lead to an increase in flooding, particularly while a channel adjusts to the differing flow regime.

5.5 Watersheds

Watersheds (or catchments, or basins) are biophysical systems that define the land surface that drains water and waterborne sediments, nutrients and chemical constituents to a point in a stream channel or a river defined by topographic boundaries (Figure 5-13).

Watershed is a geohydrological unit draining to a common point by a system of drains. All lands on the earth are part of one watershed or other. It is the land and water area, which contributes runoff to a common point. A watershed is an area of land and water bounded by a drainage divide within which the surface runoff collects and flows out of the watershed through a single outlet into a lager river (or) lake. Watersheds are the surface landscape systems that transform precipitation into water flows to streams and rivers, most of which reach the oceans. Watersheds are the systems used to study the hydrological cycle (see Chapter 2), and they help us understand how human activities influence components

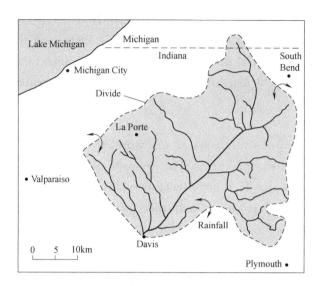

Figure 5-13 The Kankakee River Basin above Davis, IN
Source: DAVIS M L, MASTEN S J. Principles of Environmental Engineering and Science (3rd ed). 2014.

of the hydrological cycle.

5.5.1 Watersheds and Types of Drainage Pattern

Watershed approach is considered ideal for management and utilization of three basic natural resources, i.e. land, water and vegetation and their interaction in the context of watershed boundaries. It is commonly classified depending upon the size, drainage, shape and land use pattern (Table 5-2).

Classification of Watershed   Table 5-2

Classification	Area Covered(km^2)
Macro-watershed	> 500
Sub-watershed	100 to 500
Milli-watershed	10 to 100
Micro-watershed	1 to 10
Mini-watershed	< 1

Natural drainage patterns are created, where stream courses follow the lead of a landscape's geological history and features. Characteristics of the underlying rock, steepness of slope, faults and joints in the Earth's surface, specific shape of particular geological formations, and soil's susceptibility to erosion are among the factors that affect the pattern established for the flow of water in a particular place. Generally, drainage patterns are of six types. They are dendritic drainage pattern, trellis drainage pattern, radial drainage pattern, parallel drainage pattern, rectangular drainage pattern and annular drainage pattern, as shown in Figure 5-14.

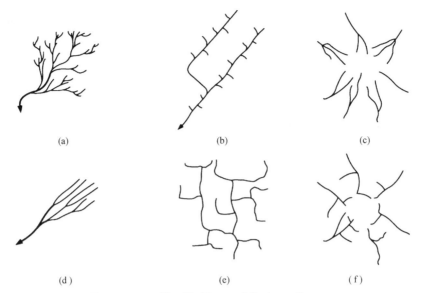

Figure 5-14 The Six Types of Drainage Patterns
(a) Dendritic Drainage Pattern; (b) Trellis Drainage Pattern; (c) Radial Drainage Pattern;
(d) Parallel Drainage Pattern; (e) Rectangular Drainage Pattern; (f) Annular Drainage Pattern

(1) Dendritic Drainage Pattern
Dendritic drainage system is the most common form of drainage system (Figure 5-14a). It has many contributing streams (analogous to the twigs of a tree), which are then joined together into the tributaries of the main river (the branches and the trunk of the tree, respectively). They develop, where the river channel follows the slope of the terrain. Dendritic drainage systems form in V-shaped valleys; as a result, the rock types must be impervious and non-porous.

(2) Trellis Drainage Pattern
The geometry of a trellis drainage system is similar to that of a common garden trellis used to grow vines (Figure 5-14b). As the river flows along a strike valley, smaller tributaries feed into it from the steep slopes on the sides of mountains. These tributaries enter the main river at approximately 90°, causing a trellis-like appearance of the drainage system. Trellis drainage is a characteristic of folded mountains.

(3) Radial Drainage Pattern
In a radial drainage system, the streams radiate outwards from a central high point (Figure 5-14c). Volcanoes usually display excellent radial drainage. Other geological features on which radial drainage commonly develops are domes and laccoliths. On these features, the drainage exhibits a combination of radial patterns.

(4) Parallel Drainage Pattern
A parallel drainage system is a pattern of rivers caused by steep slopes with some relief

(Figure 5-14d). Because of the steep slopes, the streams are swift and straight, with very few tributaries, and all flow in the same direction. This system forms on uniformly sloping surfaces.

(5) Rectangular Drainage Pattern

Rectangular drainage develops on rocks, which are of approximately uniform resistance to erosion, but have two directions of joining at approximately right angles (Figure 5-14e). The joints are usually less resistant to erosion than the bulk rock; so erosion tends to preferentially open the joints and streams eventually develop along the joints. The result is a stream system in which streams mainly consist of straight line segments with right angle bends and tributaries join larger streams at right angles.

(6) Annular Drainage Pattern

In an annular drainage pattern, streams follow a roughly circular or concentric path along a belt of weak rock, resembling a ring-like pattern (Figure 5-14f). It is best displayed by streams draining a maturely dissected structural dome or basin, where erosion has exposed rimming sedimentary strata of greatly varying degrees of hardness, which nearly encircles the domal structure.

5.5.2 Watersheds and Stream Orders

Watersheds and stream channels can be described according to their position in the landscape. It is useful to refer to an established nomenclature of stream orders in discussing watersheds and the water in streams that emanates from them. The commonly used method of stream orders classifies all unbranching stream channels as first-order streams (Figure 5-15). A second-order stream is one with two or more first-order stream channels; a third-order stream is one with two or more second-order stream channels, and so forth. Any single lower stream juncture above a larger order stream does not change the order of the larger order stream. Thus, a third-order stream that has a juncture with a

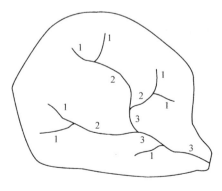

Figure 5-15 Stream Order System

second-order stream remains a third-order stream below the juncture. The watershed that feeds the stream system takes on the same order as the stream. That is, the watershed of a second-order stream is a second-order watershed and so on.

While there is little evidence that streamflow and watershed characteristics are related to stream order, the use of this terminology helps one place a stream channel or a watershed in the context of the overall drainage network of a river basin. The physical and biological characteristics of watersheds and the climate in which they exist determine the magnitude and pathways of water flow. Furthermore, the hierarchy of watersheds within a river basin generally influences the magnitude of water flow.

5.5.3 A Geomorphologic Perspective

As the upper-most watersheds in a river basin, first-order watersheds, also called headwater watersheds, are the most upstream watersheds that transform rainfall and snowmelt runoff into streamflow. Headwater streams comprise 70%~80% of total watershed areas and contribute most of the water reaching the downstream areas in river basins. Headwater watersheds are often forested or once were prior to the expansion of agriculture, urban areas and other human development activities. These headwaters are particularly important in water resource management. First-order streams in mountainous regions occur in steep terrain and flow swiftly through V-shaped valleys. High rainfall intensities can erode surface soils and generate large magnitude streamflow events with high velocities that can transport large volumes of sediment downstream.

Over geologic time, mountains erode and sediment becomes deposited downstream (Figure 5-16). As water and sediment from headwater streams merge with higher order streams, sediment is deposited over vast floodplains as rivers reach sea level. A transitional zone exists between the steep headwater streams and the lower zone of deposition at the mouth of major rivers and is typically characterized by broad valleys, gentle slopes and meandering streams.

The "work" of water on soils, hillslopes and within rivers forms landscapes with topography and soils that are better suited for some types of land use than others. Agricultural centers have developed in the transitional and depositional areas of a river basin while the steeper uplands are likely to prohibit intensive agricultural cultivation, resulting in landscapes with forests, woodlands and rangelands suitable for forestry and livestock-grazing enterprises.

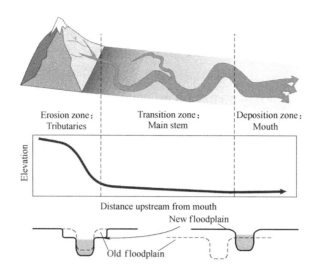

Figure 5-16 Rivers Generally Flow from an Upper, High-gradient Erosion Zone through a Transition Zone to a Low-gradient Deposition Zone

Source: BROOKS K N, FFOLLIOTT P F, MAGNER J A. Hydrology and the Management of Watersheds (4th ed). 2013.

5.5.4 A Topographic Perspective

The topography of the landscape exerts an enormous influence on the movement of water in the subsurface. Topography likewise should control the development of areas of surface saturation and runoff. If we could break a watershed up into blocks ("reservoirs"), we might be able to use the conservation of mass equation to determine the degree of saturation and potential for runoff generation for each one. Each block would differ in its position along the hillslope and in the slope of the land surface (and probably the water table) through the block. Consideration of inflows, outflows and runoff potential for all of the blocks in a watershed could provide the starting point for routing water through the watershed.

Specific contributing area can be defined for each point within a watershed if the topography of a watershed is known. Although it would be possible to estimate from a high-resolution topographic map, most studies of this sort use digital elevation data that can be used in conjunction with geographic information system (GIS) software to determine the specific contributing area for each point in the watershed (Figure 5-17). It is important to note that the results of an analysis such as this are highly dependent on the quality and resolution of the digital elevation data. Accurate identification of the channel network, in particular, depends on using high-resolution elevation data. The current state of the art technique for generating high-resolution topographic data is light detection and ranging (lidar). In fact, the critical need for good topographic data to define channels and channel

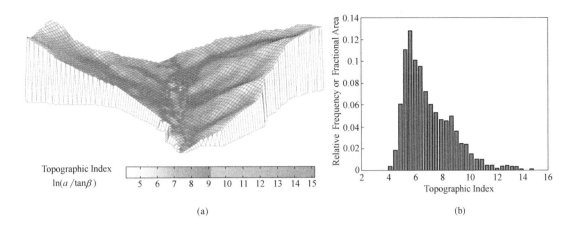

Figure 5-17 Topographic Indices for a Watershed in Shenandoah National Park
Note: The spatial pattern (a) indicates a likelihood of saturation in the central valley of the watershed.
The distribution of values (b) is calculated by using TOPMODEL.
Source: HORNBERGER G M, WIBERG P L, RAFFENSPERGER J P, et al.
Elements of Physical Hydrology (2nd ed). 2014.

networks has led to calls for lidar maps to be produced to allow accurate mapping of flood plains.

Summary

The water flowing down a river is the product of precipitation after all the other hydrological processes have been in operation. The sub-processes of overland flow, throughflow and groundwater flow are well understood, although it is not easy to estimate their relative importance for a particular site, particularly during a storm event. The measurement of river flow is relatively straightforward and presents the fewest difficulties in terms of sampling error, although there are limitations, particularly during periods of high flow and floods. Flood is a type of runoff that could cause much damage, the extent and size of which can often be related to the effect of high rainfall in the watershed.

Translation of Some Sections

部分章节参考译文

5.2 河流断面的流速分布

河流是水从高地流向低地（海洋、盆地）的主要途径。河流是城市供水、灌溉农业、工业生产和娱乐的主要水源。沉积物、营养物、重金属和其他污染物也通过河川径流向下游输送。由于自由液面和河床摩擦力的存在，速度在河流断面内的分布是不均匀的。也就是说，水流速度在水平和垂直方向上都是不同的（图 5-5）。

河流断面内的最大流速通常在靠近断面中心的自由液面以下位置。越靠近河岸方向，速度越小（图 5-5）。影响速度分布的因素是河流断面形状、通道粗糙程度和弯曲程度。地表的风对河流断面流速分布的影响很小。在实验室模拟中可观察到一种螺旋运动，但在天然河川径流中，这种螺旋运动通常很弱，其影响常被河道摩擦力抵消。

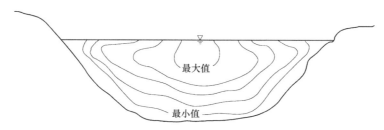

图 5-5 流速在河流断面的典型分布

来源：GUPTA R S. Hydrology and Hydraulic Systems (4th ed). 2017.

5.3 径流量的测量

径流量即穿过河流断面的水流速度。当径流量超过河道容量时，就会发生洪水。因此，有必要对径流量进行测量和预测。径流量的测量方法包括瞬时测量法和连续测量法。

5.3.1 瞬时流量测量

流速-面积法是通过测量水流的流速和水流断面面积，并将二者的值相乘得到径流量值：

$$Q = vA \tag{5-1}$$

式中　Q——径流量（m^3/s）；

　　　v——水流断面面积的平均速度（m/s）；

　　　A——断面面积（m^2）。

实际是将水流横向分成若干小断面，测量流经每个小断面的流速：

$$Q = \sum_{i=1}^{n} v_i a_i \tag{5-2}$$

式中　Q——流量（m^3/s）；

　　　v_i——在梯形 i 横截面积内测量的速度（m/s）；

　　　a_i——梯形 i 的面积（m^2）（通常按两个深度的平均值乘以两者之间的宽度估算）；

　　　n——梯形的数目。

测量流量时使用的小断面的数量取决于河床的宽度和平滑度。如果河床特别粗糙，则有必要划分出更多的小断面，使估计值尽可能接近实际值（注意图 5-6 中虚线和实线之间的差异）。

通常用一种插入水流中、记录螺旋桨随时间转动次数的螺旋桨式流量计测量水流速度（图 5-7）。在流速-面积法中假设速度测量值代表所有小断面内的平均速度。因为河床对水

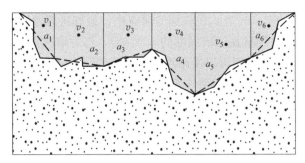

图 5-6　径流量测量的流速-面积法

注：黑色圆圈表示速度读数的位置。虚线表示三角形或梯形断面的面积，通过该横截面积测量速度。

来源：DAVIE T. Fundamentals of Hydrology (2nd ed). 2008.

流的摩擦力减缓了水流速度，所以水体表面处的速度比靠近河床处的速度快。根据经验法则，取样深度应在河流深度的60%位置（图5-8），即在1m深的河流中，取样点应在低于

图 5-7　螺旋桨式流量计

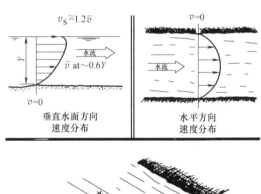

图 5-8　估算河流平均流速所需测量的河道横截面和流速

来源：BROOKS K N, FFOLLIOTT P F, MAGNER J A. Hydrology and the Management of Watersheds (4th ed). 2013.

地表 0.6m 处或高于河床 0.4m 处。在较深的河流中，最好进行两次测量（一次测量在河流深度的 20％处，另一次测量在河流深度的 80％处），然后取两次测量的平均值。

如果没有可用的流速计，则可以使用浮子对流速进行粗略的估计（即按照其移动一定距离所需的时间）。使用这种方法时须注意的是浮子在河流表面，其移动速度比河床附近的水流速度大。

流速-面积法是测量径流量的一种有效方法，其可靠性在很大程度上取决于采样策略。此方法对于浑浊且河床粗糙的小溪（如山溪），其可靠性较差。

5.3.2 连续流量测量

瞬时流量的测量仅允许在一个位置进行一次测量。虽然可在未来进行重复测量，却无法提供精确的过程数据。下面介绍 4 种连续测量方法：水位-流量关系、水槽和堰、雷达水位测量和超声波流量测量。

（1）水位-流量关系

河流水位是描述水深的术语。通过对径流量进行多次测量（即使用流速-面积法重复测量），可绘制出河流水位和流量之间的关系曲线，即流量特征曲线（图 5-9）。此法的优点是通过连续测量河流水位（操作相对简单）就可以得到实际流量。在流量-水位关系中，流量是通过在特定地点进行一系列流速-面积的测量得出的，水位来自于静水井（图 5-10）的记录。图 5-9 是苏格兰北部埃斯克河某断面的流量与水位的流量特征曲线（非线性关系），可以看出，水位较低时引起水位变化所需要的水量比水位高时所需的水量要大。

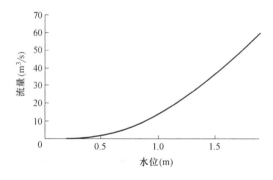

图 5-9　1963～1990 年苏格兰北部埃斯克河的流量特征曲线
来源：DAVIE T. Fundamentals of Hydrology（2nd ed）．2008．

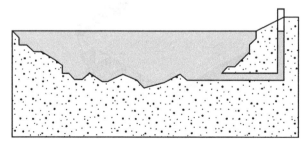

图 5-10　利用静水井连续测量河流水位（水深）
来源：DAVIE T. Fundamentals of Hydrology（2nd ed）．2008．

准确的水位-流量关系取决于河流流量测量的准确性、测量的次数,以及河床断面的稳定性。如果河床断面发生变化(例如,在大洪水期间,河床可能被冲刷或出现新的泥沙的沉积),水位-流量的关系则会发生改变,以前建立的关系将失效。为防止出现这种情况,可通过加筑混凝土结构(例如水槽或堰)的方式保持其稳定性。

流量特征曲线法的缺点是频繁测量河流流量,将使低流量和中等流量期间的测量次数多,但洪水期间的测量次数反而少。这是因为:(1) 洪水出现频率低,按照监测方案定期测量时可能错过洪水;(2) 暴发洪水期间,流量测量有较高的危险性。极端事件数据的缺乏可能导致流量特征曲线在峰值流量期间无法正确反映水位-流量关系。

(2) 水槽和堰

水槽和堰是利用上述水位-流量关系,进一步连续记录河流流量。河流流量是由一系列流经横断面的河流流速组成(如速度剖面法),那么水槽和堰或河流测量结构(图 5-11)的作用就是将这两项分离。

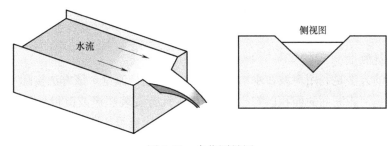

图 5-11　水位测量堰

来源:DAVIS M L, MASTEN S J. Principles of Environmental Engineering and Science (3rd ed). 2014.

第一,分离流速。通过减慢流速(或在某些情况下,加快流速),使水以恒定的速度流过一个已知的横截面。在设计水槽或堰时,不管河流水位有多高,其应能使河流以恒定的速度(或已知的速度)流过。洪水期间河流通常流速更快一些,这时就需要在测量结构之前安置一个减缓河流流速的区域:静水池。

第二,分离横断面。在水流上施加一个刚性结构,使其始终流过一个已知的横截面。这样,就可以通过测量河流水位的方法得到横截面积。除非横截面是规则的形状,河流水位一般从静水井(图 5-10)读取。

一旦速度和横断面确定下来,就可以通过试验和水力学理论得到流量特征曲线。

(3) 雷达水位测量

雷达传感器是一个具有独立单元的角状发射装置,如图 5-12 所示。它的工作原理是通过空气传播电磁波,电磁波从水面反射回到传感器。系统根据发射波和回波之间的时间间隔计算距离。传感器通过内部编程将到水面的距离转换为水位单位。数据采集平台将雷达探测到的水位数据进行存储和传输。传感器可以安装在桥或其他水面上方的合适位置。传感器有一个坚固的、相对较轻的防水外壳,且易于安装和维护。雷达传感器不占用河段内的空间,也不受河道变迁和水流中杂物碎片的影响。该装置优点是在洪水期可正常使用,缺点是不适用于结冰的河流。

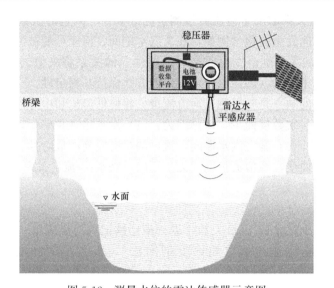

图 5-12 测量水位的雷达传感器示意图
来源：TURNIPSEED D P，SAUER V B. Discharge Measurements at Gaging Stations. 2010.

（4）超声波流量测量

当前最新的方法是利用声波在水中传播特性来测量径流量。该方法实际上测量的是水流速度，前提是河床断面的面积已知（且恒定），且有安装超声波流量计的位置。这样测得的水流速度，结合测量的水位数据，可以实现对河流流量的连续测量。有两种超声波流量计，其工作方式略有不同。

第一种方法是测量从发射器发射的超声波到达河对岸接收器所用的时间。水流速度越快，其波径的偏转就越大，其穿过河流所需的时间也就越长。声音在水中以大约 1500m/s 的速度传播（与水的纯度和深度有关），因此要求这种仪器具有高的精确性，能够在纳秒内测量。这种超声波流量计的优点是可作为永久性装置安装，缺点是河宽须大于 5m，且河流中悬浮固体含量不能过高。

第二种方法是利用多普勒效应来测量河流所携带粒子的速度。它所测量的是从悬浮粒子上反弹的超声波的波长。悬浮物粒子速度越快，超声波波长越短。这种仪器可在宽度小于 5m 的溪流中工作，同时需要一些悬浮物的参与。

5.5　流　域

流域（或分水岭、盆地）是将水和水性沉积物、营养物质和化学物质排至由地形边界定义的河道或河流中的某一点的陆地表面生物物理系统（图 5-13）。流域作为一个水文地质单元（一块陆地或水域），通过水流系统将水汇聚到一个公共点。地球上所有的土地均为某流域的一部分。一个流域就是指以排水分水岭为界的土地和水的区域。在该区域内，地表径流通过一个单一的出口汇集并流出分水岭，进入一个较宽的河流（或）湖泊。流域是将降水转化为径流的地表景观系统，其中大部分径流最终进入海洋。流域也是研究水文循环的尺度之一（见第 2 章），有助于我们了解人类活动对水文循环各组成部分的影响。

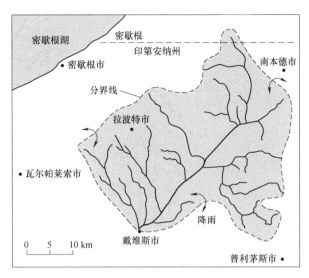

图 5-13 美国印第安纳州戴维斯市上游的坎卡基河流域

来源：DAVIS M L，MASTEN S J. Principles of Environmental Engineering and Science（3rd ed）．2014．

5.5.1 流域及水系类型

流域法是管理和利用 3 种基本自然资源的理想方法，即土地、水和植被及其在流域边界条件下的相互作用。通常根据流域的大小、排水、形状和土地利用方式对流域进行分类（表 5-2）。

流域的分类　　　　　　　　　　　　　　　　　　　　　　　　表 5-2

类　　型	覆盖面积（km²）
巨型流域	>500
中型流域	100～500
小型流域	10～100
微型流域	1～10
极小型流域	<1

自然排水模式是指河流遵循景观的地质历史和地质特征所形成的路线。影响地表水流模式的因素为下伏岩石的特性、斜坡的陡度、地表的断层和节理、地质构造的特定形状以及土壤侵蚀敏感度。一般来说，排水方式有 6 类，分别为树枝状水系、网格状水系、放射状水系、平行水系、矩形水系和环状水系（图 5-14）。

(1) 树枝状水系

树枝状排水系统是最常见的排水系统（图 5-14a）。此系统有许多支流（类似于一棵树的树枝），这些支流汇合在一起成为河流的干流（分别类似于树枝和树干），并沿着地形的斜坡在河道延伸。树枝状水系一般形成于岩石致密且不透水的 V 形山谷。

(2) 网格状水系

网格状水系的几何结构类似于用于种植葡萄的普通花园网格（图 5-14b）。当河流沿

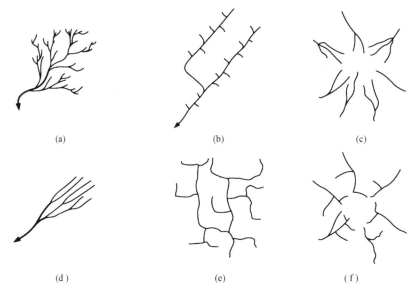

图 5-14　6 类排水方式
(a) 树枝状水系；(b) 网格状水系；(c) 放射状水系；(d) 平行水系；(e) 矩形水系；(f) 环状水系

着山谷走向流动时，较小的支流从山体两侧的陡坡流入。这些支流以大约 90°的角度进入主要河流，形成排水系统的网格状外观。网格状水系是褶皱山脉的一个特征。

(3) 放射状水系

在放射状水系中，溪流从中心高点向外辐射（图 5-14c）。火山岩一般是通过径向排水，是发育放射状水系的地质特征。穹隆和岩基也是普遍发育放射状水系的地质特征。具有这些特征的排水水系呈现出放射状的组合。

(4) 平行水系

平行排水系统是由具有一定起伏的陡坡形成的河流形态（图 5-14d）。由于坡度陡峭、溪流湍急笔直、支流极少且流向一致，此类水系易在均匀倾斜的地面上形成。

(5) 矩形水系

矩形水系发育在抗侵蚀能力一致，且两个方向的节理近似呈直角的岩石上（图 5-14e）。节理通常比大块岩石的抗侵蚀性差，因此，水流倾向于侵蚀节理，使河流最终沿着节理发育，最终使水系主要由呈直角的线段组成，且支流垂直连接干流。

(6) 环状水系

环状水系多发于成熟切割构造的穹隆或盆地，溪流沿着类似于环形模式的软岩带侵蚀而露出硬度不同的边缘沉积地层，然后形成几乎包围穹隆结构的水系（图 5-14f）。

5.5.2　流域和河流序号

流域和河道可以根据其在景观中的位置来描述。流域和由此产生的河流的描述可参照已建立的河流序号命名法。通常，将所有无支流的河道定义为一级河流（图 5-15）。二级河流是具有两个或两个以上一级河流的河流；三级河流是具有两个或两个以上二级河流的河流，依此类推。某级河流与低于此级河流的支流接合后，不会改变此级河流的级数。因此，具有与二级河流接合的接合点的三级河流仍然为三级河流。河流所在的流域与河流具有相同的级别，如二级河流的流域是二级流域。

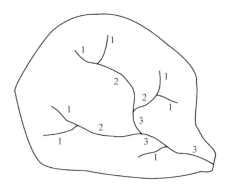

图 5-15　河流序号系统

划分流域与河流等级有助于将流域或河道置于流域整体排水网中。流域的物理、生物特征及气候决定了河流的流量和途径。此外，流域等级通常也会影响河流的流量。

5.5.3　流域地貌

一级流域作为流域中最上游的流域，又称源头流域，是将降雨和融雪转化为径流的流域。上游源头河流占流域总面积的 70%～80%，并将大部分水输送至流域的下游地区。源头流域常为森林覆盖区，或者在农业、城市发展及人类其他发展活动之前曾经是森林覆盖区。这些水源在水资源管理中尤为重要。一级河流产于山区的陡坡地带，流经 V 形河谷。高强度降雨会侵蚀表层，迅速产生大型径流，从而将大量泥沙输送到下游。

随着地质时间的推移，山脉侵蚀物和沉积物不断在下游沉积（图 5-16）。当源头河流的水和泥沙与高级别河流汇合后，随着河流到达大海，泥沙便沉积在广阔的漫滩上。在陡峭的源头河流和主要河流河口沉积带之间存在一个过渡带，其典型特征是宽阔的山谷、缓坡和曲流。

水在土壤、山坡和河流内部共同"作用"下形成了不同类型的地形和土壤景观。流域的过渡区和沉积区适于发展农业；陡峭的高地会形成森林、林地和牧场，适于林业和畜牧业，而不适于发展集约型农业。

5.5.4　流域地形

流域地形控制地表饱水区和径流区的形成，对地下水的运动有很大的影响。如果把一个流域分成几个区块（"储水区块"），可以用质量守恒定律来确定每个区块的饱和程度和产流潜力。每个区块在山坡的位置不同，其地表坡度（可能还有地下水位）也不同。若已知所有区块的径流流入量、径流流出量和径流潜力，就可以得到流域中的水路起点。

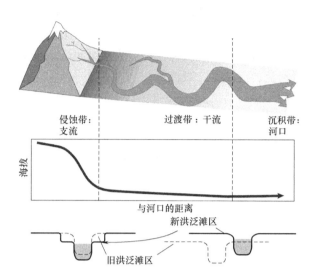

图 5-16　河流一般从较高梯度的侵蚀带流经过渡带至低梯度的沉积带
来源：BROOKS K N，FFOLLIOTT P F，MAGNER J A. Hydrology and the Management of Watersheds (4th ed). 2013.

若已知流域的地形，就可以为流域内的每个点规划特定的贡献区域。流域中每个点的具体贡献区域面积可以从高分辨率地形图中估算，但更多的是使用数字高程数据结合地理信息系统（GIS）软件来确定（图 5-17）。需要注意的是，分析结果在很大程度上取决于数字高程数据的质量和分辨率。水系网络的准确识别尤其依赖于具有高分辨率的高程数据。获取高分辨率地形数据的最新手段是光探测和测距（激光雷达）。事实上，好的地形数据对于确定河系网络十分重要，因此，准确绘制洪泛平原的激光雷达图是当前所迫切需要的。

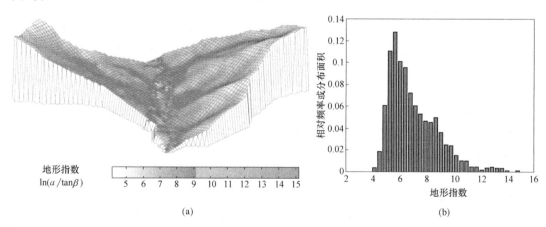

图 5-17　谢南多厄国家公园某流域的地形指数
注：图 (a) 表明了流域中央谷地饱和的可能性；图 (b) 中分布值的计算采用了 TOPMODEL。
来源：HORNBERGER G M，WIBERG P L，RAFFENSPERGER J P，et al. Elements of Physical Hydrology (2nd ed). 2014.

Questions

5-1 What is meant by streamflow and how is it determined?

5-2 What are the factors that affect streamflow?

5-3 How are natural drainage patterns created?

5-4 What ways can the topographic map of a watershed be obtained in?

Chapter 6 STREAMFLOW ANALYSIS

One of the most important tasks in hydrology is to analyze streamflow data. These data are continuous records of discharge, frequently measured in permanent structures such as flumes and weirs (Chapter 5). Analysis of these data provides us with three important features:

(1) description of a flow regime;
(2) potential for comparison between rivers;
(3) prediction of possible future river flows.

There are well-established techniques available to achieve these, although they are not universally applied in the same manner. This chapter sets out three important methods of analyzing streamflow: hydrograph analysis, flow duration curves and frequency analysis. These three techniques are explained with reference to worked examples, all drawn from the same data set. The use of data from within the same study area is important for comparison between the techniques.

6.1 Hydrograph Analysis

6.1.1 The Hydrograph

A graph of river stage or discharge versus time at a point is called a hydrograph (Figure 6-1). The former is referred to as a stage hydrograph and the latter is a discharge hydrograph. In the analysis of floods, water supply and other subjects included in surface-water hydrology, the basic quantity to be dealt with is river discharge, the rate of volume transport of water. Both river discharge and depth (or stage) change with time, and an understanding of this temporal variability is a prerequisite for approaching hydrological analyses.

The shape of a hydrograph is a response from a particular watershed to a series of unique conditions, ranging from the underlying geology and watershed shape to the antecedent wetness and storm duration. The temporal and spatial variations in these underlying conditions make it highly unlikely that two hydrographs will ever be the same. Although there is great variation in the shape of a hydrograph, there are common characteristics of a storm hydrograph that can be recognized. These have been described at the start of Chapter 5 where terms such as rising limb, recession limb and baseflow are explained (see Figure 5-2).

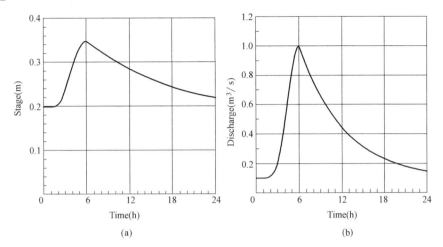

Figure 6-1 The Hydrograph
(a) River Stage as Functions of Time; (b) Discharge as Functions of Time
Source: HORNBERGER G M, WIBERG P L, RAFFENSPERGER J P, et al. Elements of Physical Hydrology (2nd ed). 2014.

6.1.2 Hydrograph Separation

The separation of a hydrograph into baseflow and stormflow is a common task, although never easy. The idea of hydrograph separation is to distinguish between stormflow and baseflow so that the amount of water resulting from a storm can be calculated. Sometimes further assumptions are made concerning where the water in each component has come from (i.e. groundwater and overland flow), but this is controversial. The simplest form of hydrograph separation is to draw a straight, level line from the point where the hydrograph starts rising until the stream discharge reaches the same level again (dashed line in Figure 6-2).

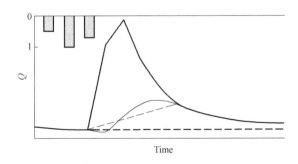

Figure 6-2 Hydrograph Separation Techniques
Source: DAVIE T. Fundamentals of Hydrology (2nd ed). 2008.

However, this is frequently problematic, as the stream may not return to its pre-storm level before another storm arrives. Equally, the storm may recharge the baseflow enough so that the level is raised after the storm (as shown in Figure 6-2). To overcome the problem of a level baseflow separation, a point has to be chosen on the receding limb where it is decided that the discharge has returned to baseflow. Exactly where this point will not be easy to determine. By convention the point is taken where the recession limb fits an exponential curve. This can be detected by plotting the natural log (ln) of discharge (Q) and noting where this line becomes straight. The line drawn between the "start" and "end" of a storm may be straight (thin dotted line in Figure 6-2) or curved (thin solid line in Figure 6-2) depending on the preference of hydrologist.

In very large watersheds, Equation (6-1) can be applied to derive the time where stormflow ends. This is the fixed time method, which gives the time from peak flow to the end of stormflow:

$$t = D^n \tag{6-1}$$

where t——the time where stormflow ends (days);

D——the drainage area (square miles);

n——a recession constant (approximately 0.2).

The problem with hydrograph separation is that the technique is highly subjective. There is no physical reason why the "end" of a storm should be when the recession limb fits an exponential curve; it is pure convention. Equally, the shape of the curve between "start" and "end" has no physical reason. What hydrograph separation does offer is a means of separating stormflow from baseflow, something that is needed for the use of the unit hydrograph, and may be useful for hydrological interpretation and description.

6.1.3 The Unit Hydrograph (UHG)

The fundamental concept of the UHG is that the shape of a storm hydrograph is determined by the physical characteristics of the watershed. The majority of those physical characteristics are static in time, therefore if you can find an average hydrograph for a particular storm size then you can use that to predict other storm events. In short: two identical rainfall events that fall on a watershed with exactly the same antecedent conditions should produce identical hydrographs.

With the UHG, a hydrologist is trying to predict a future storm hydrograph that will result from a particular storm. This is particularly useful as it gives more than just the peak runoff volume and includes the temporal variation in discharge.

A UHG is defined as "the hydrograph of surface runoff resulting from effective rainfall falling in a unit of time such as 1 hour or 1 day". The term "effective rainfall" is taken to be the rainfall that contributes to the storm hydrograph. This is often assumed to be the rainfall that does not infiltrate the soil and moves into the stream as overland flow.

(1) Deriving the UHG
Step 1: Take historical rainfall and streamflow records for a watershed and separate out a selection of typical single-peaked storm hydrographs. It is important that they are separate storms as the method assumes that one runoff event does not affect another. For each of these storm events, separate the baseflow from the stormflow; that is, hydrograph separation. This will give you a series of storm hydrographs (without the baseflow component) for a corresponding rainfall event.

Step 2: Take a single storm hydrograph and find out the total volume of water that contributes to the storm. This can be done either by measuring the area under the stormflow hydrograph or as an integral of the curve. If you then divide the total volume in the storm by the watershed area, you have the runoff as a water equivalent depth. If this is assumed to have occurred uniformly over space and time within the watershed, then you can assume it is equal to the effective rainfall. This is an important assumption of the method: that the effective rainfall is equal to the water equivalent depth of storm runoff. It is also assumed that the effective rainfall all occur during the height of the storm (i.e. during the period of highest rainfall intensity). That period of high rainfall intensity becomes the time for the UHG.

Step 3: The UHG is the stormflow that results from one unit of effective rainfall. To derive this you need to divide the values of stormflow (i.e. each value on the storm hydrograph) by the amount of effective rainfall (from step 2) to give the UHG. This is the discharge per millimeter of effective rainfall during the time unit.

Step 4: Repeat steps 2 and 3 for all of the typical hydrographs. Then create an average UHG by merging the curves together. This is achieved by averaging the value of stormflow for each unit of time for every derived UHG.

(2) Using the UHG
The UHG obtained from the steps described here theoretically gives you the runoff that can be expected per millimeter of effective rainfall in one hour. In order to use the UHG for predicting a storm, it is necessary to estimate the "effective rainfall" that will result from the storm rainfall. This is not an easy task and is one of the main hurdles in using the method. In deriving the UHG, the assumption has been made that "effective rainfall" is the rainfall, which does not infiltrate but is routed to the stream as overland flow. The

same assumption has to be made when utilizing the UHG. To do this, it is necessary to have some indication of the infiltration characteristics for the watershed concerned, and of the antecedent soil moisture conditions. The former can be achieved through field experimentation and the latter using an antecedent precipitation index (API). The idea is that antecedent soil moisture is controlled by how long ago rain has fallen and how large that event was. The wetter a watershed is prior to a storm, the more effective rainfall will be produced.

Once the effective rainfall has been established, it is a relatively simple task to add the resultant UHG together to form the resultant storm hydrograph. The worked example shows how this procedure is carried out.

The rainfall quantity and distribution over time from a design storm is obtained initially for the watershed. Estimated loss rates are then subtracted from total rainfall to obtain the quantity, distribution and duration of effective rainfall. The rainfall duration is divided by the selected UHG duration to obtain the number of periods to be added up for the total design storm. The ordinates of the selected UHG are then multiplied by the quantity of effective rainfall for each period. In Table 6-1, for example, 20mm of effective rainfall yields a stormflow hydrograph with ordinates 20 times those of the corresponding UHG for the first period and 30mm of effective rainfall is 30 times the UHG but delayed for the second period. The calculated stormflows plus any baseflow are summed for each period to obtain the total stormflow hydrograph.

Application of a 1-hour UHG to a storm of 2-hour of effective precipitation   Table 6-1

Time(h)	1-hour UHG Ordinates (m^3/s)	Effective Rainfall (mm)	Streamflow			Baseflow (m^3/s)	Total Discharge (m^3/s)
			Time 1 (m^3/s)	Time 2 (m^3/s)	Subtotal (m^3/s)		
0	0	0	0	0	0	1.2	1.2
1	0.05	20	1	0	1	1.2	2.2
2	0.5	30	10	1.5	11.5	1.2	12.7
3	1	0	20	15	35	1.2	36.2
4	0.75	0	15	30	45	1.2	46.2
5	0.5	0	10	22.5	32.5	1.2	33.7
6	0.25	0	5	15	20	1.2	21.2
7	0	0	0	7.5	7.5	1.2	8.7
8	0	0	0	0	0	1.2	1.2

The assumption of linearity is not always valid. As effective rainfall increases, the magnitude of the peak actually can increase more than the proportional increase in the rainfall amount. A consequence of ignoring a nonlinear response can underestimate the magnitude of peak discharge for large storm events. If a nonlinear response is suspected, two or more UHGs should be developed from observed hydrographs resulting from substantially different rainfall amounts. The appropriate UHG would then be applied only to precipitation amounts similar to those used to develop the UHG in the first place.

(3) Limitations of the UHG

The UHG has several assumptions that at first appearance would seem to make it inapplicable in many situations. The assumptions can be summarized as:

1) The runoff that makes up stormflow is derived from infiltration excess (Hortonian) overland flow. This is not a reasonable assumption to make in many areas of this planet.

2) The surface runoff occurs uniformly over the watershed because the rainfall is uniform over the watershed. This is another assumption that is difficult to justify.

3) The relationship between effective rainfall and surface runoff does not vary with time (i.e. the hydrograph shape remains the same between the data period of derivation and prediction). This would assume no land-use change within the watershed, as this could well affect the storm hydrograph shape.

Given the assumptions listed above, it would seem extremely foolhardy to use the UHG as a predictive tool. However, the UHG has been used successfully for many years in numerous different hydrological situations. It is a very simple method of deriving a storm hydrograph from a relatively small amount of data. The fact is that it does work (i.e. produces meaningful predictions of storm hydrographs), despite being theoretically flawed. The UHG is a black box model of stormflow and as such hides many different processes within. The simple concept that the hydrograph shape is a reflection of the static characteristics and all the dynamic processes going on in a watershed makes it highly applicable but less able to be explained in terms of hydrological theory.

6.2 Flow Duration Curves

An understanding of how much water is flowing down a river is fundamental to hydrology. Of particular interest for both flood and low flow, hydrology is the question of how representative a certain flow is. This can be addressed by looking at the frequency of daily flows and some statistics that can be derived from the frequency analysis. The culmination of the frequency analysis is a flow duration curve, which is described below.

Flow duration curves are concerned with the amount of time of a certain flow which is exceeded. The data most commonly used are daily mean flows: the average flow for each day (this is not the same as a mean daily flow, which is the average of a series of daily flows). To derive a flow duration curve, the daily mean flow data are required for a long period of time, in excess of five years.

(1) Deriving Flow Duration Curve

Step 1: A table is derived that has the frequency, cumulative frequency and percentage cumulative frequency. The percentage cumulative frequency is assumed to equal the percentage of time that the flow is exceeded. While carrying out the frequency analysis, it is important that a small class (or bin) interval is used; If the interval is too large, information will be lost from the flow duration curve. The method for choosing the best class interval is essentially through trial and error. Generally, you should aim not to have more than around 10% of your values within a single class interval. If you have more than this, you start to lose precision in plotting.

Step 2: The actual flow duration curve is created by plotting the percentage cumulative frequency on the x-axis against the mid-point of the class interval on the y-axis (Figure 6-3). If two flow duration curves were to be presented on the same axes, they need to be standardized before direct comparison. To do this, the values on the y-axis (mid-point of class interval) are divided by the average flow for the record length. The presentation of a flow duration curve may be improved by either plotting on a special type of graph paper or transforming the data. A natural log transformation of the flow values (y-axis) achieves a similar effect, although this is not necessarily standard practice (Figure 6-4).

(2) Interpreting a Flow Duration Curve

The shape of a flow duration curve can tell a lot about the hydrological regime of a watershed. In Figure 6-5, two flow duration curves of contrasting shape are shown. With the dotted line, there is high variability in flow, whereas for the solid line there is far less

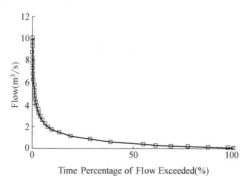

Figure 6-3　Flow Duration Curve for the river Wye

Source: data from the upper reaches of the river Wye in Mid-Wales, UK (1970~1995).

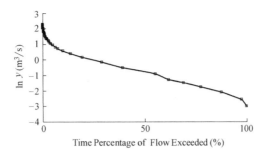

Figure 6-4 Flow Duration Curve with the Data Same as in Figure 6-3
Shown on a Natural Log Scale

variation. This tells us that the watershed shown by the solid line never has particularly low flows or particularly high flows, and its baseflow is high. This type of hydrological response is found in limestone or chalk watersheds where there is a high baseflow in the summer (groundwater derived) and high infiltration rates during storm events. In contrast, the watershed shown with a dotted line has far more variation. During dry periods, it has a very low flow, but responds to rainfall events with a high flow. This is characteristic of impermeable upland watersheds or streamflow in dryland areas.

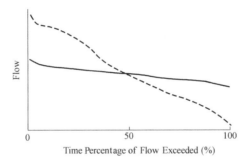

Figure 6-5 Two Contrasting Flow Duration Curves

(3) Statistics Derived from a Flow Duration Curve

The interpretation of flow duration curve shape discussed so far is essentially subjective. In order to introduce some objectivity, there are statistics derived from the curve; the three most important ones are:

1) The flow value that is exceeded 95% of the time (Q95), which is a useful statistic for low flow analysis;

2) The flow value that is exceeded 50% of the time (Q50), which is the median flow value.

3) The flow value that is exceeded 10% of the time (Q10), which is a useful statistic for analysis of high flows and flooding.

6.3 Frequency Analysis

The analysis of how often an event is likely to occur is an important concept in hydrology. It is the application of statistical theory into an area that affects many people's lives, whether it is through flooding or low flows and drought. Both of these are considered here, although because they use similar techniques, the main emphasis is on flood frequency analysis. It is an attempt to place a probability on the likelihood of a certain event occurring. Predominantly it is concerned with the low frequency, high-magnitude events (e.g. a large flood or a very low river flow).

It is important to differentiate between the uses of flow duration curves and frequency analysis. Flow duration curves tell us the percentage of time that a flow is above or below a certain level. This is average data and describes the overall flow regime. Flood frequency analysis is concerned only with peak flows: the probability of a certain flood recurring. Conversely, low flow frequency analysis is concerned purely with the lowest flows and the probability of them recurring.

Flood frequency analysis is concerned with peak flows. There are two different ways that a peak flow can be defined:

1) the single maximum peak within a year of record giving an annual maximum series;

2) any flow above a certain threshold value, giving a partial duration series.

6.3.1 Frequency Distribution

The first step in reaching flood frequency distribution is to obtain the data series (annual maxima discharge). The annual maximum series should be for as long as the data record allows. The greater the length of record is, the more certainty can be attached to the prediction of average recurrence interval. There is an assumption made in flood frequency analysis that the peak flows are independent of each other (i.e. they are not part of the same storm). If a calendar year is chosen for a humid temperate environment in the northern hemisphere, or a tropical region, it is possible that the maximum river flow will occur in the transition between years. It is possible for a storm to last over the 31 December/1 January period and the same storm to be the maximum flow value for both years. Therefore, if the flow regime is dominated by snowmelt, it is important to avoid splitting the hydrological year at times of high melt (e.g. spring and early summer). To avoid this, it is necessary to choose your hydrological year as changing during the period of lowest flow. This may take some initial investigation of the data.

The second step in reaching flood frequency distribution is to obtain the frequency histogram and probability distribution. It is often useful to convert the frequency into a relative frequency.

On looking at the exampled histogram of the data set (Figure 6-6), the first obvious point to note is that it is not normally distributed (i.e., it is not a classic bell-shaped curve). It is important to grasp the significance of the non-normal distribution for two reasons:

1) Common statistical techniques that require normally distributed data (e.g. t-tests, etc.) cannot be applied in flood frequency analysis.

2) It shows what you might expect: small events are more common than large floods, but that very large flood events do occur (i.e. a high magnitude, low frequency relationship).

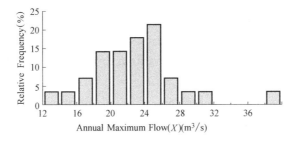

Figure 6-6 Frequency Distribution of the Annual Maximum Series
Source: DAVIE T. Fundamentals of Hydrology (2nd ed). 2008.

If you were to assume that the data series is infinitely large in number and the class intervals were made extremely small, then a smooth curve can be drawn through the histogram. This is the probability density function, which represents the smoothed version of your frequency histogram.

6.3.2 Flood Frequency Analysis

In flood frequency analysis, there are three interrelated terms of interest. These terms are interrelated mathematically, as described in Equation (6-2) and Equation (6-3) in the text below.

(1) The probability of excedence: $P(X)$. This is the probability that a flow (Q) is greater than, or equal to a value X. The probability is normally expressed as a unitary percentage (i.e. on a scale between 0 and 1).

(2) The relative frequency: $F(X)$. This is the probability of the flow (Q) being less than a value X. This is also expressed as a unitary percentage.

(3) The average recurrence interval (or return period): $T(X)$. This is sometimes referred to as the return period, although this is misleading. $T(X)$ is a statistical term meaning the chance of exceedance once every T years over a long record. This should not be interpreted as meaning that is exactly how many years are likely between certain size floods.

$$P(X)=1-F(X) \tag{6-2}$$

$$T(X)=\frac{1}{P(X)}=\frac{1}{1-F(X)} \tag{6-3}$$

It is possible to read the values of $F(X)$ from a cumulative probability curve; this provides the simplest method of carrying out flood frequency analyses. One difficulty with using this method is that you must choose the class intervals for the histogram carefully so that the probability density function is an accurate representation of the data. One way of avoiding the difficulties of choosing the best class interval is to use a rank order distribution. This is often referred to as a plotting position formula——The Weibull Formula.

The first step in the method is to rank your annual maximum series data from low to high. In doing this, you are assuming that each data point (i. e. the maximum flood event for a particular year) is independent of any others. This means that the year that the flood occurred in becomes irrelevant. Taking the rank value, the next step is to calculate the $F(X)$ term using Equation (6-4). In this case, r refers to the rank of an individual flood event (X) within the data series and N is the total number of data points (i. e. the number of years of record):

$$F(X)=\frac{r}{N+1} \tag{6-4}$$

In applying the formula, there are two important points to note:
① The value of $F(X)$ can never reach 1 (i. e. it is asymptotic towards the value 1).
② If you rank your data from high to low (i. e. the other way around) then you will be calculating the $P(X)$ value rather than $F(X)$. This is easily rectified by using the formula linking the two.

The Weibull Formula is simple to use and effective but is not always the best description of an annual maximum series data. Some users suggest that a better fit to the data is provided by the Gringorten Formula (Equation 6-5):

$$F(X)=\frac{r-0.44}{N+0.12} \tag{6-5}$$

As illustrated in the worked example, the difference between these two formulae is not great and often the use of either one is down to personal preference.

Summary

The analysis of streamflow records is extremely important in order to characterize the flow regime for a particular river. Hydrograph analysis involves dissecting a hydrograph to distinguish between stormflow and baseflow. This is often a precursor to using the UHG, a technique using past stormflow records to make predictions on the likely form of future storm events. Flow duration curves are used to look at the overall hydrology of a river——the percentage of time that a river has an average flow above or below a certain threshold. Frequency analysis is used to look at the average return period of a rare event (or the probability of a certain rare event occurring), whether that is extremes of flooding or low flow. Each of the methods described in this chapter has a distinct use in hydrology and it is important that practicing hydrologists are aware of their role.

Translation of Some Sections

部分章节参考译文

6.1 过程线分析

6.1.1 过程线

河流水位或流量随时间变化的曲线称为过程线（图6-1）。前者称为水位过程线，后者称为流量过程线。在分析洪水、供水等地表水文学问题时，要处理的基本量是河流流量、水量输运率。了解河流流量和水位（或水深）随时间的变化情况是进行水文分析的先决条件。

过程线的形状是特定流域对一系列独特基本条件的响应，这些基本条件包括流域地质、流域形状、先前的湿度和暴雨持续时间。不同的流域基本条件造就了不同形状的过程线。尽管过程线形状不同，但它们具有一些共同的特征（详见第5.1节），如涨水段、退水段和基流等（图5-2）。

6.1.2 过程线分离

水文工作者的一项常见任务是将过程线分离为基流和地表径流。过程线分离的目的是计算降水产生的地表径流量。有时会对水来自何处（即地下和地表）作进一步假设，但是这是有争议的。最简单的过程线分离法是从过程线开始上升的点处绘制一条直线，直至流

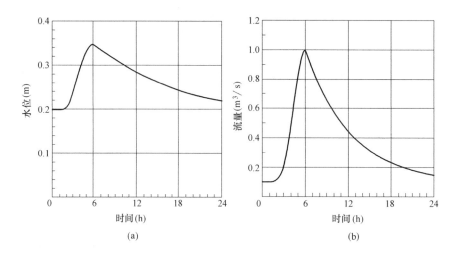

图 6-1 过程线
(a) 河流水位随时间变化线；(b) 流量随时间变化线
来源：HORNBERGER G M，WIBERG P L，RAFFENSPERGER J P，et al.
Elements of Physical Hydrology（2nd ed）. 2014.

量再次达到此相同的水平（图 6-2 中的虚线）。

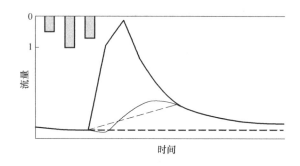

图 6-2 过程线分离方法
来源：DAVIE T. Fundamentals of Hydrology（2nd ed）. 2008.

但问题是，在下一场暴风雨来之前，河流可能还没有恢复到降雨前的水位。同样，降雨也可能会对基流产生足够的补给，这样降水过后基流水位将提高（如图 6-2 所示）。为了克服水平分离基流的缺陷，可在退水段上确定一个点，并认为该点上的流量已返回基流。但是，确定这一点位置并不容易。按照惯例，这一点是指衰退段边缘符合指数曲线处，可以通过绘制流量的自然对数曲线与流量曲线的切点位置来确定。在降水"开始"和"结束"之间画的线可以是直线（图 6-2 所示细虚线）或曲线（图 6-2 所示细实线），这些是基于水文工作者的主观偏好。

在较大的流域中，可以采用固定时间法（式 6-1）来计算从洪峰开始到降水结束的时间：

$$t = D^n \tag{6-1}$$

式中　t——暴雨结束的时间（天）；

D——流域面积（平方英里）；

n——衰退常数（约 0.2）。

过程线分离过程是高度主观的。将衰退边缘符合指数曲线时的那一点当作降雨的"结束"点，并不是基于物理上的推理，而是纯粹地基于惯例。同样，"开始"和"结束"点之间的曲线形状也没有经过物理推理。但是，过程线分离确实提供了一种将降水流量与基流分离的方法，这是利用单位过程线进行水文分析的基础，并且可用于水文解释和水文描述。

6.1.3 单位线（UHG）

单位线的基本概念是：降雨过程线的形状由流域的物理特征决定，这些物理特征大多数是静态的，因此，如果可以找到某级降雨的平均过程线，则可以使用该过程线来预测其他降雨事件。简而言之，两个相同的降雨落在一个具有完全相同的流域上，其产生的过程线应该是相同的。

有了单位线，就可以预测指定的降雨产生的流量过程线。流量过程线可体现出峰值流量和流量随时间的变化情况。

单位线的定义为："单位时间内（如 1h 或 1 天）有效降雨量引发的地表径流过程"。"有效降雨量"指的是产生流量过程线的降雨量，即有效降雨不渗入土壤而形成径流的那部分降雨量。

(1) 单位线的推导

步骤 1：利用流域的历史降雨量和流量分离确定出典型的单峰径流过程线。因为该方法假定降雨事件不会互相影响，因此要保证降雨事件的独立性。对这些降雨事件进行过程线分离，为相应的降雨事件提供一系列不包括基流的流量过程线。

步骤 2：用流量过程线计算降雨量。这可以通过测量流量过程线下的面积或对曲线作积分来实现。将总降雨量除以流域面积，则可得到径流的当量深度。假设 1：降雨在流域的空间和时间上均匀发生，则可以假设它等于有效降雨量，即有效降雨量等于地表径流的当量深度。假设 2：有效降雨量均发生在最高降雨强度期间，且这一时间段为单位线的时间。

步骤 3：单位线是指一单位有效降雨量产生的径流过程。将流量过程线上的每个值除以有效降雨量（步骤 2 得到的），即可得出单位线。

步骤 4：对所有典型的过程线重复步骤 2 和 3。然后通过合并曲线创建一个平均单位线（对每个单位线的每个单位时间的流量值作平均）。

(2) 单位线的使用

从理论上讲，单位线给出了 1h 内每毫米有效降雨量产生的预期径流。为了利用单位线预测降雨，需要估算"有效降雨量"，实际做起来并不容易。在推导单位线时，单位线中涉及的"有效降雨量"是指不发生渗透而形成地表径流的那部分降雨的量。因此需要考察有关流域的入渗特性和降雨前土壤的含水情况。前者可以通过田间试验来获取，后者则是利用前期降水指数（API）来表征。API 的理念是指降雨前期土壤湿度由此次降雨与上次降雨的时间间隔和上次降雨的规模控制。降雨前土壤越湿润，有效降雨量越大。

一旦确定了有效降雨量，就可以将生成的单位线叠加在一起生成径流过程线，具体过程如下例所示：

已知流域设计降雨的降雨量和其随时间的分布。从总降雨量中减去损失的量，得到有效降雨量、降雨分布和持续时间。降雨持续时间除以所选用的单位线的持续时间，以获得总设计降雨的累积周期数。然后，用选定的单位线坐标乘以每个时段的有效降雨量。例如，在表 6-1 中，20mm 的有效降雨量产生的径流过程线的纵坐标是第 1 个时段对应单位线的 20 倍，30mm 的有效降雨量是第 2 个时段对应单位线的 30 倍，注意此时有一个时段的延迟。计算出的径流流量叠加每个时段的基流可得到总的径流流量过程线。

2h 有效降雨产生 1h 单位线的应用实例　　　　　　　　表 6-1

时间(h)	1h UHG 纵坐标 (m^3/s)	有效降雨量 (mm)	各时段降雨的地面径流			基流 (m^3/s)	总径流 (m^3/s)
			第 1 时段 (m^3/s)	第 2 时段 (m^3/s)	合计 (m^3/s)		
0	0	0	0	0	0	1.2	1.2
1	0.05	20	1	0	1	1.2	2.2
2	0.5	30	10	1.5	11.5	1.2	12.7
3	1	0	20	15	35	1.2	36.2
4	0.75	0	15	30	45	1.2	46.2
5	0.5	0	10	22.5	32.5	1.2	33.7
6	0.25	0	5	15	20	1.2	21.2
7	0	0	0	7.5	7.5	1.2	8.7
8	0	0	0	0	0	1.2	1.2

然而，线性叠加并不总是成立的。随着有效降雨量的增加，实际峰值会比按降雨量线性叠加的值更高。忽略非线性响应的后果可能会低估降雨事件的洪峰流量。如果非线性影响较大，应基于降雨量不同测量值产生的流量过程线推导出两个或多个单位线，最终选用与最初用于推导单位线的降水量相似的降水量所推导出的单位线。

（3）单位线的局限性

单位线的几个假设在许多情况下不适用。这些假设可以概括为：

1）地表径流来源于超渗产流（即霍顿）。这个假设在很多地方无法适用。

2）地表径流在流域内均匀分布是由于流域内降雨量均匀分布。这也是一个难以证明的假设。

3）有效降雨量和地表径流之间的关系不随时间变化（即在推导和预测数据期间，过程线形状保持不变）。这假定流域内的土地利用类型没有发生变化，因为土地利用类型很可能会影响径流过程线的形状。

单位线法是一种从相对较少的数据中导出流量过程线的简单方法。虽然有很多缺陷和不足，但确实可以产生有意义的流量过程线。其多年来已成功地应用于许多不同的水文情况。单位线法隐藏了许多不同的过程，可以说是降雨-径流的黑箱模型，是静态特征和流域内所有动态过程的共同反映。虽然单位线的形状难以用水文理论来解释，但它仍具有很强的实际应用价值。

6.2 流量历时曲线

径流量是水文学的重要研究方向。研究河流丰枯水时，主要考察的是某流量值具有多高的代表性。流量历时曲线是利用对河流日流量的统计数据绘制而成的。

流量历时曲线与超过某一流量值的时间有关。最常用的数据是日平均流量（这与平均日流量不同，平均日流量是一系列日流量的平均值）。绘制流量历时曲线一般需要至少5年内的日平均流量数据。

(1) 流量历时曲线的推导

步骤1：列出包含频率、累积频率和累积频率百分比的表格。假设累积频率百分比等于超过流量值的时间所占的百分比。进行频率分析时，区间要尽可能的小，以避免信息丢失。选择最佳区间的方法即为试错法。通常，在一个区间内的值相差不超过10%。如果超过10%，曲线的精度就会下降。

步骤2：以累积频率百分比为横坐标，以流量区间的中点为纵坐标，绘制流量历时曲线（图6-3）。如果要在同一坐标上显示两条流量历时曲线，则需先进行标准化处理，即横坐标不变，纵坐标除以记录范围内的平均流量。流量历时曲线也可以在特殊图表上绘制或通过转换数据来改进。对流量值取自然对数（lny）可达到类似的效果，但这并非标准步骤（图6-4）。

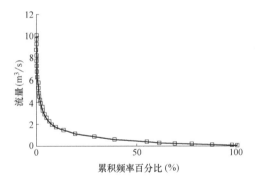

图6-3 怀伊河流量历时曲线

来源：数据来源于1970~1995年英国威尔士中部的怀伊河流量统计。

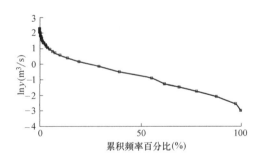

图6-4 对图6-3数据取自然对数后的流量历时曲线

(2) 流量历时曲线的解读

流量历时曲线的形状可以很好地反映流域的水文状况。图6-5显示了两条不同形状的流量历时曲线。用虚线表示的流量历时曲线，流量变化较大；而用实线表示的流量历时曲线，流量变化较小。这说明，实线所代表的流域无特别低或特别高的流量，而且其基流量较高。这种类型的水文响应一般出现在石灰岩流域。石灰岩流域通常在夏季有较高的基流（地下充沛），且暴雨期间有较高的渗透率。相比之下，虚线所代表的流域其流量变化较大。在干旱时期，其流量很低，但对降雨事件的响应却很高。这是不透水的高地流域或

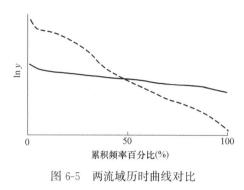

图 6-5 两流域历时曲线对比

干旱区河流的典型特征。

(3) 从流量历时曲线得出的统计数据

目前,对流量历时曲线形状的解释是较主观的。为了加强客观性,可从曲线获得一些统计数据,其中最重要的 3 个统计数据是:

1) 超过所考察时间段 95% 的流量值($Q95$),可用于低流量的分析;

2) 超过所考察时间段 50% 的流量值($Q50$),即中间流量值;

3) 超过所考察时间段 10% 的流量值($Q10$),可用于高流量和洪水的分析。

6.3 频率分析

分析水文事件发生的频率是水文学研究中的一项重要课题,也是影响人们生活领域(不管洪水还是干旱)的一项统计学应用。在这里,我们重点关注的是洪水频率分析,主要涉及罕见事件(如大洪水或大干旱)。

流量历时曲线和频率分析的应用范围不同。流量历时曲线可以体现出流量高于或低于某值的时间所占的百分比。这时的流量是平均数据。洪水频率分析只涉及洪峰流量,利用其来获取某一级洪水的重现概率。相反,干旱频率分析只关注最低流量及其重复出现的概率。

在洪水频率分析中,有两种不同的方法可定义洪峰流量:

1) 记录一年内的最大单峰,给出一个年最大序列;

2) 高于某一阈值的任何流量,给出部分持续时间序列。

6.3.1 频率分布

进行洪水频率分布分析的首要步骤是获取数据序列(一般是年最大流量序列)。年最大值系列的记录时长应尽可能长一些。记录的时间跨度越大,预测平均重现期的可靠性就越高。洪水频率分析是基于洪峰流量彼此独立的假设。在为北半球的潮湿温带地区或热带地区选择一个水文年的时候,有可能最大年径流量出现在跨年时段。例如一场风暴可能从 12 月 31 日持续至第二年的 1 月 1 日,也就是说同一场风暴均为这两年的最大流量值。因此,需要对数据进行初步调查。若在融雪期(如春季和初夏),则应选择在最低流量期划分水文年。

进行洪水频率分布分析的第 2 个步骤是绘制频率直方图和概率分布图。通常将频率转换为相对频率。

频率直方图(图 6-6)是非正态分布的,需要正确把握其非正态分布的意义,因为:

1) 正态分布数据的统计方法(如 t 检验等)不能应用于洪水频率分析。

2) 非正态分布显示了可能的预期:小事件比大洪水更常见,但非常大的洪水事件(即高级别、低频率的)也是会发生的。

假设数据序列在数量上无限大,并且区间非常小,则可基于直方图绘制平滑曲线,即

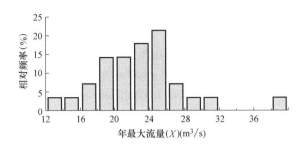

图 6-6 年最大流量序列频率分布

来源：DAVIE T. Fundamentals of Hydrology (2nd ed). 2008.

为概率密度函数曲线。

6.3.2 洪水频率分析

在洪水频率分析中，有 3 个相关的术语。这些术语在数学上是相互关联的，其关系如式（6-2）和式（6-3）所示：

$$P(X)=1-F(X) \tag{6-2}$$

$$T(X)=\frac{1}{P(X)}=\frac{1}{1-F(X)} \tag{6-3}$$

（1）超越概率：$P(X)$ 是指流量大于或等于 X 值的概率，通常用百分比表示（即介于 0 和 1 之间）。

（2）相对频率：$F(X)$ 是指流量小于 X 值的概率，也用百分比表示。

（3）平均重现间隔（或重现期）：$T(X)$ 表示在长期记录中每 T 年发生一次超过数的概率，是一个统计术语，但它并不表示某级别的洪水每隔 T 年就会发生一次。

$F(X)$ 的值可以从累积概率曲线中读取，这是进行洪水频率分析的最简单方法。使用此方法时，对直方图间距的选取决定着概率密度函数精确程度。最佳的间距选择方法是秩次分布法，即利用 Weibull 公式。

首先，将年度最大序列数据从低到高排序。在此过程中，假设每个数据点（即特定年份的最大洪水事件）相互独立。这意味着洪水发生的年份各不相关。接下来，利用式（6-4）计算 $F(X)$ 项。在这种情况下，r 是指数据系列中单个洪水事件（X）的等级（即秩），N 是数据点的总数（即记录年数）：

$$F(X)=\frac{r}{N+1} \tag{6-4}$$

应用此公式时需要注意：

① $F(X)$ 的值永远小于 1（但它可以趋近于 1）。

② 若将数据从高到低（即相反）排序，则得到的是 $P(X)$ 值而不是 $F(X)$。

Weibull 公式的使用虽简单、有效，但并不总是适合描述年最大序列。一些观点认为 Gringorten 公式（式 6-5）将更适合：

$$F(X)=\frac{r-0.44}{N+0.12} \tag{6-5}$$

如前述示例所示，这两个公式之间的形式差别不大，可使用其中任何一个。

Questions

6-1 Find a scientific paper in the literature that uses a hydrological model, and evaluate the type of model and its strengths and weaknesses for the study concerned.

6-2 Outline the limitations of the unit hydrograph when used as a predictive tool and attempt to explain its success despite these limitations.

6-3 Describe the types of information that can be derived from a flow duration curve and explain the use of that information in hydrology.

6-4 Explain why interpretation of flood (or low flow) frequency analysis may be fraught with difficulty.

6-5 Explain what is meant by a 100-year flood.

6-6 From the following hourly streamflow record (Table 6-2) due to a storm, separate the baseflow by the recession curve technique. The drainage area is 30 acres. Also, determine the runoff depth (cfs means cubic feet per second).

Hourly Streamflow Record   Table 6-2

Time	Flow(cfs)	Time	Flow(cfs)
1	30	9	45
2	29.4	10	31.5
3	66	11	22.5
4	155	12	18
5	190	13	16
6	140	14	14.5
7	100	15	13
8	63		

6-7 The peak-flow data on an annual basis from Cedar River near Austin, Minnesota State, are listed in Table 6-3. Plot the flood-frequency curve on lognormal probability paper. Determine: (a) magnitude of a flood having a return period of 100 years (probability of 1%), and (b) probability of a flow of 100 cfs (cfs means cubic feet per second).

6-8 The baseflow in a stream and the 3-hour unit hydrograph for the basin are given below (Figure 6-7 and Table 6-4). Determine the total flow hydrograph for a storm of the pattern indicated.

Chapter 6 STREAMFLOW ANALYSIS

Peak-flow Data on an Annual Basis from Cedar River Table 6-3

Year	Peak Flow (cfs)	Year	Peak Flow (cfs)	Year	Peak Flow (cfs)	Year	Peak Flow (cfs)
1965	7750	1978	979	1991	3880	2003	8690
1966	5440	1979	4940	1992	2110	2004	3410
1967	3580	1980	4260	1993	8270	2005	2190
1968	5260	1981	9400	1994	5740	2006	4440
1969	4000	1982	9530	1995	4140	2007	4070
1970	8800	1983	2330	1996	3910	2008	1400
1971	7070	1984	990	1997	1300	2009	3290
1972	7520	1985	9400	1998	12400	2010	7580
1973	6990	1986	3740	1999	4720	2011	4640
1974	5570	1987	3250	2000	3250	2012	3190
1975	2710	1988	2920	2001	4810	2013	10800
1976	2190	1989	3830	2002	3060	2014	2500
1977	2250	1990					

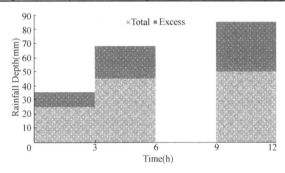

Figure 6-7

Baseflow and the 3-hour Unit Hydrograph Table 6-4

Time	Unit Hydrograph(m^3/s)	Baseflow(m^3/s)
12:00	0	10
15:00	4.7	10
18:00	7.6	11
21:00	5.9	11
24:00	4.3	11
3:00	3.1	12
6:00	2.4	11
9:00	1.3	12
12:00	0.7	13
15:00	0.2	12
18:00	0	13

111

Chapter 7 GROUNDWATER HYDROLOGY

Water in saturated zones lying beneath the soil surface is groundwater. Groundwater comprises more than 97% of all liquid freshwater on Earth. More than one-half of the world population depends on groundwater. Although groundwater is an important source of freshwater, it does not always occur where it is most needed and is sometimes difficult to extract.

From ancient times until the 1900s, the focus of groundwater science has been finding and developing groundwater resources. Groundwater is still a key resource and it always will be. In some places, it is the only source of fresh water. In the 20th century, engineering and environmental aspects of groundwater have also become important.

7.1 Porous Materials

The voids or open spaces between the particles of soils and other granular materials are referred to as pores. An important property of water-bearing material is their porosity. Porosity, the total void space between the grains and in the cracks of aquifers and solution cavities that can fill with water, is defined in terms of the percentage of pore space as

$$n = 100 \frac{V_v}{V_t} \tag{7-1}$$

where n——the porosity;
 V_v——the volume of void space in the porous material;
 V_t——the total volume of porous material including void space.

The porosity of a soil depends primarily on its particle size distribution and on its structure. Some of these features are illustrated in Figure 7-1 and Figure 7-2. A soil with a wide distribution of particle sizes tends to have a smaller porosity than a soil consisting of particles or grains of a more uniform size. The structure of a granular porous material refers to the arrangement of the particles among one another and to their aggregation into larger structures. Thus, the porosity can be increased by agricultural operations, such as ploughing or raking, or by frost; these processes "open up" the soil simply by rearranging the relative positions of the particles. Similarly, the porosity of the soil can be decreased by compaction. In principle, in the case of soils consisting of inert materials, their texture, that is the size of the particles, should not affect the porosity, as long as their structure, particle size distributions and chemical composition are similar. However, actual soils are not inert, but the surfaces of their particles carry electrical charges; these charges affect the structure of the soil, and are increasingly effective with decreasing particle size. Moreover, their chemical composition can vary widely. Some of the soil constituents, such as colloidal clay, organic matter, lime, and colloidal oxides of iron and aluminum, can act as cementing agents, which further the aggregation of particles into larger structures. As a result, clayey soils tend to have higher porosities than sandy soils.

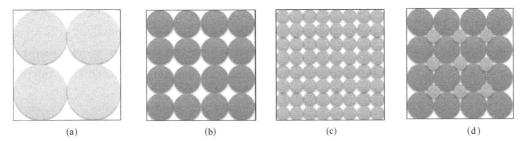

Figure 7-1 Illustration of the Effect of Texture and of Particle Size Distribution on the Porosity

Note: In boxes (a), (b) and (c), the spherical particles of uniform size arranged in a cubic packing (i. e. a similar structure) result in aggregates with exactly the same porosity, regardless of the sizes of the particles. The particles in box (d), which are of different sizes, result in an aggregate with smaller porosity.

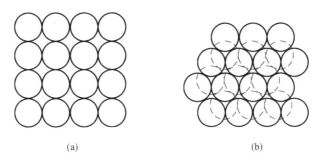

Figure 7-2 Illustration of the Effect of Structure of Spherical Particles of Uniform Size on Porosity

Note: Among the regular arrangements, the cubic packing in (a) is the most open and the rhombohedral packing in (b) is the tightest.

The effective porosity is the ratio of the void space through which water can flow to the total volume. The pores must be connected to each other if water is to move through a soil or rock stratum. If the pores are interconnected and sufficient in size to allow water to move freely, the soil or rock is permeable. Aquifers that contain small pores or weakly connected pores yield only small amounts of water even if their total porosity is high.

7.2 Aquifer Classification

7.2.1 Aquifers and Aquitards

Groundwater is found in many different types of soil and rock strata. It can occur between individual soil or rock particles, in rock fracture openings, and in solution openings formed when water dissolves mineral constituents in the rock strata, leaving a void. The amount of groundwater stored and released from the water-bearing strata depends on the porosity, the size of pore spaces and the continuity of pores. Water-bearing porous soil or rock strata that yield significant amounts of water to wells are called aquifers. It is a layer of unconsolidated or consolidated rock that is able to transmit and store enough water for extraction. An aquifer can be from a meter to hundreds of meters thick and can underlie a few hectares or thousands of square kilometers.

Slowly permeable geologic stratas that retard the movement of groundwater such as lacustrine clays, fractured mudstone and certain sandstones are aquitards. An aquitard is a geological formation that transmits water at a much slower rate than the aquifer. This is an oddly loose definition, but reflects the fact that an aquitard only becomes so relative to an aquifer. The aquitard becomes so because it is confining the flow over an aquifer. In another place, the same geological formation may be considered as an aquifer. The term aquifuge is sometimes used to refer to a totally impermeable rock formation (i.e. it could never be considered as an aquifer).

7.2.2 Aquifer Characteristics

When considering the development of groundwater for pumping, certain characteristics of the aquifer(s) from which the groundwater is to be extracted need to be understood. An important characteristic is the transmissivity of an aquifer, which is the amount of water that can flow horizontally through the saturated thickness of the aquifer under a hydraulic gradient of 1 m/m. Transmissivity is defined as

$$T_r = bK \tag{7-2}$$

where T_r——the transmissivity (m^2/h);

  b——the saturated thickness (m);

  K——the hydraulic conductivity of the aquifer (m/h).

Any time the hydraulic head in a saturated aquifer changes, groundwater will be either stored or discharged. Storativity is the volume of water that is either stored or discharged from a saturated aquifer per unit surface area per unit change in head. The storativity characteristic of an aquifer is related to the specific yield of the soil or rock material that constitutes the aquifer. Specific yield is the ratio of the volume of water that can drain freely from saturated earth material due to the force of gravity to the total volume of the earth material.

7.2.3 Unconfined or Confined Aquifer

Aquifers are sources of usable groundwater. They are classified as either unconfined or confined. Aquifers that contain water in direct contact with the atmosphere through porous material are called unconfined aquifers. The groundwater system illustrated in Figure 7-3 is unconfined with the soil system immediately above the water table, allowing the exchange of gases and water. Unconfined aquifers could be rapidly recharged by water that percolates downward from the land surface. The water table occurs within the aquifer layer, therefore, unconfined aquifers are also called water table or phreatic aquifers.

Confined aquifers are capped by an impermeable layer of rock or clay, which can cause water pressure to build up underground. In a confined aquifer, the whole thickness of the aquifer layer is saturated and there is an impermeable confining layer (aquitard) above (Figure 7-4). The water level in a stilling well in a confined aquifer rises above the top of

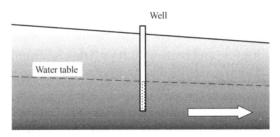

Figure 7-3 An Unconfined Aquifer
Note: The water level in the well is at the water table.
Source: DAVIE T. Fundamentals of Hydrology (2nd ed). 2008.

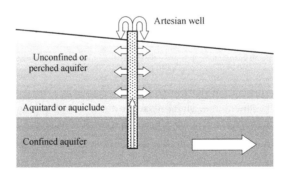

Figure 7-4 A Confined Aquifer
Note: The height of water in the well will depend on the amount of pressure within the confined aquifer.
Source: DAVIE T. Fundamentals of Hydrology (2nd ed). 2008.

the aquifer. The perched water table is an imaginary surface called the potentiometric surface. It is defined by the heads measured in wells in a confined aquifer. In plan view, the potentiometric surface is a contour map showing the horizontal distribution of heads in a confined aquifer. An unconfined aquifer can become a confined aquifer at some distance from the recharge area.

The lower boundary of the unconfined aquifer may be impervious, but it is the upper boundary, or water table, that is unconfined and may intersect the surface. It has no boundary above it, and therefore, the water table is free to rise and fall dependent on the amount of water contained in the aquifer. The level that water rises up to from a confined aquifer is dependent on the amount of fall (or hydraulic head) occurring within the aquifer (Figure 7-4). Sometimes, the water level in a well in a confined aquifer will rise above the level of the ground surface and flow freely without pumping. Artesian wells are most common at the base of slopes in hilly terrain, where the high heads under the uplands can induce strong upward hydraulic gradients.

7.3 Storage of Groundwater

All water beneath the surface is groundwater, and it is convenient to distinguish between the saturated and unsaturated zones. As shown in Figure 7-5, there is movement of water through both vertical infiltration and horizontal flow (this is a combined vector effect). It is important to realize that this occurs in both the unsaturated and saturated zones, although at a slower rate in the unsaturated.

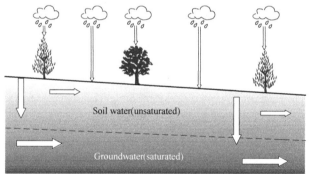

Figure 7-5 Water Stored Beneath the Earth's Surface

Note: Rainfall infiltrates through the unsaturated zone towards the saturated zone.
The broken line represents the water table, although, as the diagram indicates,
this is actually a gradual transition from unsaturated to fully saturated.
Source: DAVIE T. Fundamentals of Hydrology (2nd ed). 2008.

7.3.1 Water in the Unsaturated Zone

In some heterogeneous settings, there may be perched aquifers: zones of saturation completely surrounded by unsaturated zones, as illustrated in Figure 7-6. Layers of aquitard or aquifuge always form the base of the perched zones. If the layer is extensive, the body of perched water may be thick enough to allow a water supply well to tap it without drilling deeper to the regional water table, as shown in Figure 7-6.

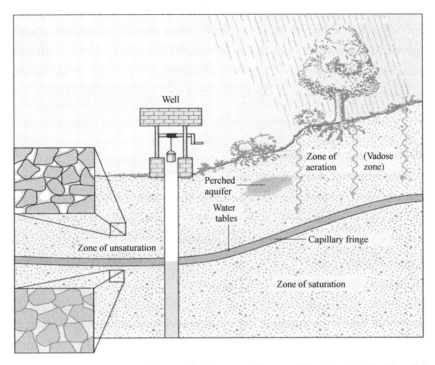

Figure 7-6　Profile of the Unsaturated Zone, the Saturated Zone and the Perched Aquifer of Groundwater
Source: DAVIS M L, MASTEN S J. Principles of Environmental Engineering and Science (3rd ed). 2014.

The majority of water in the unsaturated zone is held in soil. Soil is essentially a continuum of solid particles (minerals, organic matter), water and air. Consideration of water in the soil starts with the control over how much water enters a soil during a certain time interval: the infiltration rate. The rate at which water enters a soil is dependent on the current water content of the soil and the ability of a soil to transmit the water. Soil water content is normally expressed as a volumetric soil moisture content or soil moisture fraction (Equation 7-3):

$$\theta = \frac{V_w}{V_t} \qquad (7\text{-}3)$$

where　θ——the soil moisture content (m^3/m^3);

　　　V_w——the volume of water in a soil sample;

　　　V_t——the total volume of soil sample.

Soil water content may also be described by gravimetric soil moisture content or percentage of saturated. Gravimetric soil moisture content is the ratio of the weight of water in a soil to the overall weight of the soil. Saturated water content is the maximum amount of water that the soil can hold. Soil water content as percentage of saturated is a useful method of telling how wet the soil actually is.

Porosity is another important soil water property. It is the fraction of pore space in the total volume of soil (Equation 7-1). In theory, water can fill all of the pores in a soil; therefore, porosity is the maximum potential volumetric water content. In practice, the volumetric soil moisture seldom reaches the porosity value and if it does, gravity acts on the water to force drainage through the profile that quickly drops moisture levels back below porosity.

7.3.2 Water in the Saturated Zone

Once water has infiltrated through the unsaturated zone, it reaches the water table and becomes groundwater. This water moves slowly and is not available for evaporation (except through transpiration in deep-rooted plants), consequently it has a long residence time. This may be so long as to provide groundwater reserves available from more pluvial (i.e. greater precipitation) times. This can be seen in the Middle East, for example, Saudi Arabia is able to draw on extensive "fossil water" reserves.

However, it would be wrong to think that all groundwater moves slowly; it is common to see substantial movement of the water and regular replenishment during wetter months. In limestone areas the groundwater can move as underground rivers, although it may take a long time for water to reach these conduits. In terms of surface hydrology, groundwater plays an important part in sustaining streamflows during summer season.

7.4 Surface Tension and Capillarity

Because polar water molecules are attracted to each other, a mass of water has internal cohesion that tends to hold it together. This is apparent in the way water drops tend to form spheres as they fall through the air and the way water can bead up on a flat surface. At an air-water interface, it looks as though there is a thin membrane stretched over the surface, hence the term surface tension.

Water beads up to a greater extent on some surfaces than on others. For example, it beads up more on a freshly waxed car than on an unwaxed car. Water molecules have very little attraction to the molecules in wax, which are nonpolar. By contrast, water's self-attraction is quite strong, and water pulls itself into distinct beads on the waxy surface. If you place water drops on most any rock surface, the water will not bead up much. Instead, it will spread out and wet the surface. In pore spaces containing both air and water, the water will generally wet the mineral surfaces, leaving the central parts of the pores for the air, as illustrated in Figure 7-7.

The attraction of water for mineral surfaces causes water to pull and spread itself across the surfaces. A layer of water molecules approximately 0.1 to 0.5μm thick is so strongly attracted to the mineral surfaces that it is essentially immobile. Further than about 0.5μm from mineral surfaces, the forces of attraction are not strong enough to prevent movement of water molecules and water outside that distance is free to move. Surface attraction forces are stronger for clay minerals than for other mineral types, due to the charged nature of clay mineral surfaces.

Because of this pull or tension in the water, the pressure within the water is less than the air pressure within the pores. As the amount of water present decreases, the pull of the mineral surface attraction forces increases, the pressure within the water decreases, and the air-water interface develops a more contorted shape conforming to the mineral grains. This attraction of water for the mineral surfaces in partly saturated materials is called capillarity. Capillarity allows water to wet pore spaces above the water table, in the same way that a paper towel will wick up water when dipped into a puddle. Capillary forces tend to be greater in finer-grained granular materials, due to a greater amount of mineral surface area.

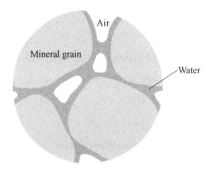

Figure 7-7 Water and Air in the Pore Spaces of a Granular Medium
Source: FITTS C R. Groundwater Science. 2002.

7.5 Movement of Groundwater

7.5.1 Water Transmit in Pores

The ability of a soil to transmit water is dependent on the pore sizes within it and most importantly on the connections between pores. A well-structured soil consists of stable aggregates with a wide range of pore sizes within and between the aggregates. In this case macropores may make up at least 10% of this soil volume.

This structure provides numerous interconnected pathways for the flow of water with a wide range of velocities. Biological activity (e.g. roots and worms) can produce macropores that provide flow paths for water that are largely separated from the main soil matrix. These are essentially two different types of macropores: those are large pores within the soil matrix; and those are essentially separated from the matrix.

(1) Infiltration

Infiltration is the net movement of water into soil. When the rainfall rate exceeds the infiltration rate, water migrates through the surface soil at a rate that generally decreases with time until it reaches a constant value. The rate of infiltration varies with rainfall intensity, soil type, surface condition and vegetal cover. This temporal decline in the rate is actually due to a filling of the soil pores with water and a reduction in capillary action. The change in infiltration rate with time is shown in Figure 7-8. When the rate of rainfall exceeds the rate of infiltration, we can use Horton's Equation (Equation 7-4).

$$f = f_c + (f_0 - f_c)e^{-kt} \tag{7-4}$$

where f——infiltration rate (or capacity);

f_c——equilibrium, critical or final infiltration rate;

f_0——initial infiltration rate;

k——empirical constant;

t——time.

Note that this equation has the same problem as many of the other mass-balance equations used by hydrologists, that is, rates are given in units of length per unit time rather than mass per unit time or volume per unit time.

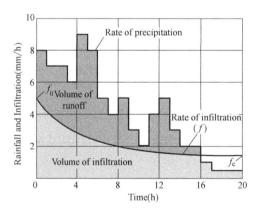

Figure 7-8 Typical Infiltration Curve——Effect of Infiltration Rate with Time

As mentioned earlier, soil type affects the infiltration rate of water. As one might imagine, the sandier the soil is, the greater the infiltration rate will be. The more compacted the soil is, or the greater the clay content of the soil is, the slower the infiltration rate will be. Table 7-1 provides data for some common soil types. Horton's infiltration can be integrated to yield an equation that represents the total volume of water that would infiltrate over a given period. The integrated form of Horton's Equation is given in Equation (7-5).

$$V = A_s \int_0^t f \, dt = A_s \int_0^t f_c + (f_0 - f_c) e^{-kt} \, dt = A_s \left[f_c t + \frac{f_0 - f_c}{t} (1 - e^{-kt}) \right] \quad (7\text{-}5)$$

Parameters for Horton's Equation for Some Typical Soils　　　　Table 7-1

Soil Type	f_c (cm/h)	f_0 (cm/h)	k (1/h)
Alphalpha Loamy Sand	3.56	48.26	38.29
Carnegie Sandy Loamy	4.5	35.52	19.64
Dothan Loamy Sand	6.68	8.81	1.4
Fuquay Pebbly Loamy Sand	6.15	15.85	4.7
Leefield Loamy Sand	4.39	28.8	7.7
Troop Sand	4.57	58.45	32.71

Source: BEDIENT P B, HUBER W C, VIEUX B E. Hydrology and Floodplain Analysis (4th ed). 2008.

Although Horton's Equation is generally applicable to most soils, several limitations apply. For sandy soils, f_0 exceeds most rainfall intensities. In these cases, Horton's Equation will underestimate the infiltration rate. As mentioned previously, the infiltration capacity f decreases with cumulative infiltration volume as the pores fill up, not with time. Note that in the Horton's Equation, infiltration capacity is a function of time, not cumulative infiltration volume.

(2) Hydraulic Conductivity

Rocks or soils with small pores allow only slow migration of water while materials with larger, less constricted pores permit more rapid migration. Water traveling through small, constricted pores must shear itself more in the process of traveling a given distance than water traveling through larger pores. More shearing in the water causes more viscous resistance and slower flow. Other factors being equal, the average velocity of groundwater migration is proportional to hydraulic conductivity (K). Hydraulic conductivity is an empirical constant measured in laboratory or field experiments.

Table 7-2 lists some typical ranges of hydraulic conductivity values for common rocks and soils. Hydraulic conductivity varies over a tremendous range (16 orders of magnitude) and in common geologic materials.

The wide variations of fracture width and frequency in crystalline rocks account for the huge ranges in observed hydraulic conductivities in such rock. Where carbonate rocks have been eroded by dissolution, fractures widen to form large openings and talk of "underground rivers" is not just mythology. Some basalts are also very conductive due to open columnar joints and voids at the bases and tops of successive lava flows. Groundwater flow velocities in basalts and limestone can be extremely high compared to velocities in more typical geologic materials where the pore sizes are on the order of millimeters or smaller.

Typical Values of Hydraulic Conductivity　　　　　　　　　　　　Table 7-2

Material	K (cm/s)
Gravel	10^{-1} to 100
Clean Sand	10^{-4} to 1
Silty Sand	10^{-5} to 10^{-1}
Silt	10^{-7} to 10^{-3}
Glacial Till	10^{-10} to 10^{-4}
Clay	10^{-10} to 10^{-6}
Limestone and Dolomite	10^{-7} to 1
Fractured Basalt	10^{-5} to 1
Sandstone	10^{-8} to 10^{-3}
Igneous and Metamorphic Rock	10^{-11} to 10^{-2}
Shale	10^{-14} to 10^{-8}

7.5.2 Principles of Groundwater Flow

In almost any investigation involving groundwater, questions arise about how much water is moving and how fast it is flowing. Typical questions in contamination remediation studies are: "What should the well discharge be to capture the entire plume of contaminated water?" or "How long will it take for the contaminated groundwater to reach a nearby stream?"

The answers to such questions are based on groundwater flow analyses, which in turn are based on some straightforward physical principles that govern subsurface flow. An empirical relationship called Darcy's Law and conservation of mass form the basis for many useful hand calculations and computer simulations that can be made to analyze groundwater flow.

In 1856, French engineer Henry Darcy was working for the city of Dijon, France on a project involving the use of sand to filter the water supply. He performed laboratory experiments to examine the factors that govern the rate of water flow through sand. The results of his experiments defined basic empirical principles of groundwater flow that are embodied in an equation now known as Darcy's Law.

Darcy's apparatus consisted of a sand-filled column with an inlet and an outlet similar to that illustrated in Figure 7-9. Two manometers (essentially very small piezometers) measure the hydraulic head at two points within the column (h_1 and h_2). The sample is saturated, and a steady flow of water is forced through at a discharge rate Q [L^3/T].

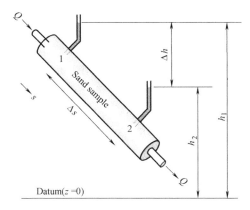

Figure 7-9 Schematic Illustrating Steady Flow through a Sand Sample
Note: The manometers measure heads h_1 and h_2 at locations 1 and 2 within the column.
The s coordinate direction runs parallel to the column.
Source: FITTS C R. Groundwater Science. 2002.

Darcy found through repeated experiments with a specific sand that Q was proportional to the head difference Δh between the two manometers and inversely proportional to (∞) the distance between manometers Δs:

$$Q \propto \Delta h, Q \propto \frac{1}{\Delta s} \tag{7-6}$$

Obviously, Q is also proportional to the cross-sectional area of the column A. Combining these observations and writing an equation in differential form gives Darcy's Law for one-dimensional flow:

$$Q_s = -K_s \frac{dh}{ds} A \tag{7-7}$$

where Q_s——discharge from an aquifer in the s direction;
   K_s——the hydraulic conductivity in the s direction;
 dh/ds——the hydraulic gradient;
   A——the cross-sectional area.

The minus sign on the right side of this equation is necessary because head decreases in the direction of flow. If there is flow in the positive s direction, Q_s is positive and dh/ds is negative. Conversely, when flow is in the negative s direction, Q_s is negative and dh/ds is positive. The h term in the hydraulic gradient includes both the elevation and pressure head. The head pressure h and the coordinate s both have length units, so dh/ds is dimensionless. The dimensionless quantity dh/ds represents the rate that head changes in the s direction, and is known as the hydraulic gradient. In an unconfined aquifer, it can be assumed that the hydraulic gradient is equal to the drop in height of water table over a horizontal distance (i.e. the elevation head). In a confined aquifer, it is the drop in phreatic surface over a horizontal distance.

[**Example 7-1**] A sample of silty sand is tested in a laboratory experiment just like that illustrated in Figure 7-9. The column has an inside diameter of 10cm and the length between manometers is $\Delta s=25$cm. With a steady flow of $Q=1.7$cm^3/min, the head difference between the manometers is $\Delta h=15$cm. Calculate the hydraulic conductivity K_s.

[**Solution**] This is a direct application of Equation (7-7), with a little rearrangement at the beginning to isolate K_s:

$$K_s = -Q_s \frac{1}{A} \frac{ds}{dh}$$
$$= -1.7 \cdot \frac{1}{\pi \cdot 5^2} \frac{25}{-15}$$
$$= 0.036 \text{cm/min}$$
$$= 6.0 \times 10^{-4} \text{cm/s}$$

The sign of ds/dh is negative because as s increases, h decreases (Figure 7-9).

7.5.3 Aquifer Recharge and Discharge

Groundwater is typically on the move. It interacts directly with surface water through aquifer recharge and discharge. The interaction between groundwater and streamflow is complex and depends very much on local circumstances. Water naturally moves towards areas where faster flow is available and consequently can be drawn upwards towards a stream. This is the case in dry environments but is dependent on there being an unconfined aquifer near the surface. If this is not the case, then the stream may be contributing water to the ground through infiltration. Figure 7-10 shows two different circumstances of interaction between the groundwater and stream. In Figure 7-10(a), the groundwater is contributing water to the streamflow as the water table is high. In Figure 7-10(b), the water table is low and the stream is contributing water to the groundwater. This is commonly the case where the main river source may be mountains a considerable distance away and the river flows over an alluvial plain with the regional groundwater table considerably deeper than stream level.

Groundwater discharges from the saturated zone back to the ground surface in low-lying areas, usually at springs or the bottom of surface waters. Since groundwater always moves towards lower head, these exit points are always at a lower elevation than the water table where groundwater enters the system as recharge.

Figure 7-11 illustrates a hypothetical vertical cross-section showing how groundwater

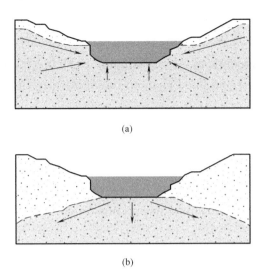

Figure 7-10 The Interactions between a River and the Groundwater
(a) The Groundwater is Contributing to the Stream; (b) The Opposite Situation to (a) is Occurring
Source: DAVIE T. Fundamentals of Hydrology (2nd ed). 2008.

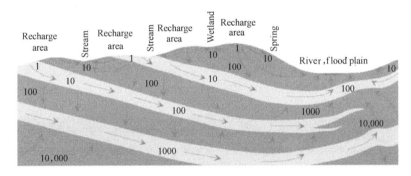

Figure 7-11 Vertical Cross-section Showing Groundwater Recharge and Discharge Areas
in a Hypothetical Setting
Note: Arrows show the direction of flow, and numbers indicate the residence time of groundwater in years. Lighter shading indicates aquifers and darker shading indicates aquitards.
Source: FITTS C R. Groundwater Science. 2002.

moves from recharge areas to discharge areas. Recharge enters the groundwater system under upland areas. It then migrates towards discharge areas at low spots in the topography. These are often the sites of wetlands, springs, streams and lakes.

The path-lines that water travels are often very irregular due to the heterogeneous distribution of hydraulic conductivities in the subsurface. In a high-conductivity layer, water tends to flow parallel to the layer boundaries. But in a low-conductivity layer, water tends to take the shortest path through the layer, flowing nearly perpendicular to the layer boundaries.

At shallow depth, there are many small, localized flow patterns as shown in Figure 7-11. Deeper down, groundwater tends to flow in larger regional patterns that reflect the larger scale geology and topography. For example, in Figure 7-11, the deeper flow is focused towards the low river and flood plain on the right side of the profile.

The residence time for groundwater (elapsed time since infiltrating) can range from days to hundreds of thousands of years. Longer residence times are found in larger groundwater basins and in deeper parts of the flow system, where there are long flow paths between recharge areas and discharge areas and slower velocities. Water in aquifers moves much more rapidly than water in aquitards, as illustrated by the residence times posted in Figure 7-11.

Numerous mathematical model simulations were conducted to demonstrate the effect of water table configuration and heterogeneity on regional groundwater flow patterns. The models were of two-dimensional steady flow in the vertical plane. Results of two of their models are shown in Figure 7-12. In each model, the water table undulates up and down creating local flow patterns at shallow depth. The subsurface is heterogeneous with the light-shaded areas 100 times more pemeable than the dark-shaded areas. The models demonstrate that localized flow systems occur at shallow depth and that heterogeneity can have a huge impact on flow patterns.

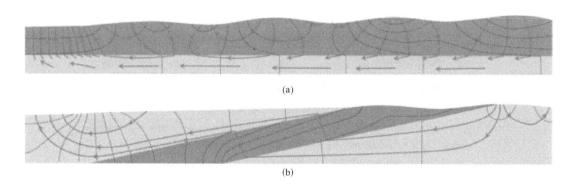

Figure 7-12　Two Mathematical Models of Steady Flow in the Vertical Plane
Source: FREEZE R A, WITHERSPOON P A. Theoretical analysis of regional groundwater flow: Effect of water-table configuration and subsurface permeability variation. Water Resources Research, 1967, 3(2): 623-634.

A spring is a place where groundwater discharges up to the ground surface. At a spring, the water table intersects the ground surface. Springs commonly occur near the base of a steep slope. Many are located where fractures or the base of an aquifer intersects the slope. Springs in several settings are illustrated in Figure 7-13.

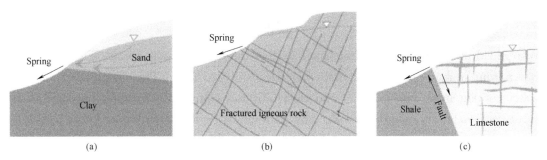

Figure 7-13 Springs in Different Geologic Settings
Source: FITTS C R. Groundwater Science. 2002.

In some regions, the water table is relatively deep below the ground surface, only intersecting the surface at major stream channels. This situation occurs when the recharge rates are low relative to the transmissivity of the near-surface materials. Figure 7-14(a) shows such a setting. The water table is deep and gently sloping, and there are long flow paths from recharge areas to the major streams. Streams and wetlands are widely spaced in such settings.

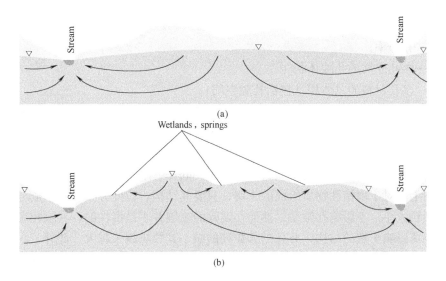

Figure 7-14 The Water Table Position where the Recharge/Transmissivity Ratio
is Low (top) and High (bottom)
Source: FITTS C R. Groundwater Science. 2002.

Figure 7-14(b) shows the opposite situation. The recharge rates are high relative to the transmissivity. The water table is much higher, just under the surface and intersecting the surface at numerous topographic depressions. Low areas form soggy wetlands, springs, ponds and streams. There are many relatively short, local flow paths in such settings.

Summary

Water held underground is an important part of the global water storage. The release of water from storage may have a significant effect on river flows. This chapter begins with water content in porous material and then introduces physical principles: properties of subsurface materials and groundwater flow. Groundwater flow according to Darcy's Law is described, and the general flow equations are derived for confined and unconfined aquifers. The recharge and discharge theories are addressed as well.

Translation of Some Sections
部分章节参考译文

7.1 多孔介质

土壤颗粒和其他颗粒之间的空隙称为孔隙。含水材料的一个重要性质是孔隙率。孔隙率，即颗粒之间可充水空间，可以定义为孔隙所占的百分比：

$$n = 100 \frac{V_v}{V_t} \tag{7-1}$$

式中　　n——孔隙率；

　　　　V_v——多孔介质的孔隙体积；

　　　　V_t——多孔材料的总体积（包括孔隙空间）。

土壤的孔隙率主要取决于其粒径分选情况和结构，其中一些特性如图 7-1 和图 7-2 所示。颗粒粒径分选差的土壤往往比分选好的土壤孔隙率要小。多孔颗粒材料的结构是指颗粒的排列方式。因此，土壤孔隙率可以通过农耕作业（如犁或耙）或霜冻来增加。这些过程通过重新排列颗粒的相对位置来"打开"土壤。同样，土壤的孔隙率也可以通过压实来降低。原则上，对于由惰性物质组成的土壤，只要其结构、粒度分布和化学成分相似，其质地（即颗粒大小）不影响孔隙率。然而，实际的土壤并不是惰性的，其颗粒表面带有电荷。这些电荷影响土壤的结构，并且随着粒径的减小其影响越来越大。此外，它们的化学成分可以有很大的不同。一些土壤成分（如胶状黏土、有机质、石灰以及铁和铝的胶状氧

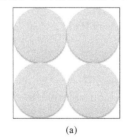　　　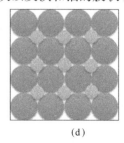

　　　(a)　　　　　　　　　　(b)　　　　　　　　　　(c)　　　　　　　　　　(d)

图 7-1　排列方式和粒度分布对孔隙率的影响

注：在图 (a)、(b) 和 (c) 中，以立方填料（即类似结构）排列，均匀尺寸的球形颗粒产生完全相同的孔隙率，与颗粒的大小无关。图 (d) 中的颗粒大小不同，导致其产生较小的孔隙率。

化物）可以作为胶粘剂，进一步将颗粒聚集成更大的结构。因此，黏性土往往比砂土具有更高的孔隙率。

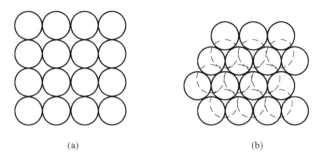

图 7-2　均匀尺寸球形颗粒结构对孔隙率的影响
注：在规则排列中，图（a）的立方堆积最为开放，图（b）的菱形堆积最为紧密。

有效孔隙率是水可以流进的空隙空间与多孔材料总体积的比值。若水要移动通过土壤或岩层，孔隙必须相互连接。如果孔隙相互连接，并且有足够的尺寸允许水自由流动，则土壤或岩石称为可渗透的。含小孔隙或弱连通孔隙的含水层，即使其总孔隙率很高，也只能渗出少量的水。

7.2　含水层分类

7.2.1　含水层和隔水层

地下水存在于许多不同类型的土壤和岩层空隙中。空隙可能是土壤或岩石颗粒之间的孔隙、岩石裂隙以及水溶解岩层中的矿物成分而形成的溶隙。含水层储存和释放地下水的量取决于孔隙率、空隙空间大小和空隙的连续性。含水量大的多孔土壤或岩层称为含水层。它是一层松散或固结的岩石，能够传输和储存足够的水用于开采。含水层的厚度从1m到数百米不等，在地下可以有几公顷或数千平方公里的面积。

渗透缓慢的地质地层，如湖相黏土、裂缝性泥岩和某些砂岩，称为隔水层。它是一种传输水的速度较慢（比含水层慢得多）的地质构造。这个定义并不十分严格。事实上，隔水层的这种性质是相对于含水层来说的。隔水层限制了含水层的水流，但在另一个地方，隔水层有可能被视为含水层。"完全不透水层"一词是指完全不透水的岩层（其无论何时均不可视为含水层）。

7.2.2　含水层特征

开采地下水时，需要了解所要开采的含水层的特征。含水层的一个重要的特征是其输水性能，即在1m/m的水力梯度下，水平流过饱和含水层的水量，其按下式求解：

$$T_r = bK \tag{7-2}$$

式中　T_r——输水性能；

b——饱和含水层的厚度；

K——含水层的导水率。

饱和含水层的水头发生变化时，地下水将被储存或排出。释水系数是指单位表面积内单位水头变化时从饱和含水层中储存或排出的水量。含水层的储存特性与构成含水层的土壤或岩石材料的单位产水量有关。单位产水量是由于重力作用而从饱和土壤中自由排出的水量与土壤或岩石总体积的比值。

7.2.3 非承压和承压含水层

人类可利用的地下水来自于含水层。含水层可分为非承压含水层和承压含水层。通过多孔介质与大气直接接触的含水层称为非承压含水层。图 7-3 所示的非承压地下水不受正上方土壤的约束，允许气体和水进行交换。非承压含水层可通过地表水向下渗透而迅速得到补给。地下水水位线在非承压含水层内，因此，非承压地下水也称为潜水。

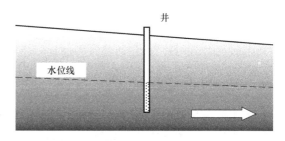

图 7-3 非承压含水层

注：井里的水位即地下水位。

来源：DAVIE T. Fundamentals of Hydrology（2nd ed）. 2008.

承压含水层被不可渗透的岩石或黏土层所覆盖，导致地下水压累积。在承压含水层中，整个含水层厚度是饱和的，上面有一个不透水层（隔水层，图 7-4）。承压水的水位是一个假想的势能面，可通过承压含水层静水井中测得的水头来确定。在平面图中，势能面表现为承压含水层中水平分布的水头等高线。非承压含水层在距离补给区一定距离处可视为承压含水层。

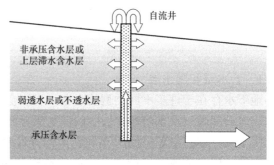

图 7-4 承压含水层

注：井内水位的高低取决于承压含水层内的压力大小。

来源：DAVIE T. Fundamentals of Hydrology（2nd ed）. 2008.

非承压含水层的下边界一般是不透水的，而它的上边界或地下水位不承受压力，因此地下水位随水量自由升降，可与地表相交。承压含水层中水可以上升到的高度取决于承压含水层的水头（图 7-4）。当承压含水层的水位高于地表，井水可自由溢出，这种承压含水层称为自流含水层，该井称为自流井。在丘陵地区，自流井最常见于坡底，因为高地下方的高水头可引起强烈的向上的水力梯度。

7.3 地下水的存储

地表以下的水均为地下水，分为饱和带水和非饱和带水。如图 7-5 所示，水的运动是一种由垂直入渗和水平流动组合的矢量效应。非饱和区与饱和区均存在水的运动，且在非饱和区水的速度较慢。

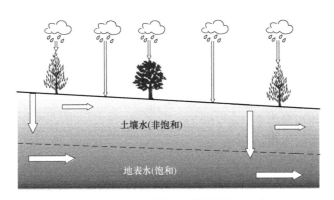

图 7-5 地球表面下储存的水

注：虚线表示地下水位，降雨通过非饱和带向饱和带渗透，即地下水从不饱和到完全饱和的逐渐过渡。

来源：DAVIE T. Fundamentals of Hydrology (2nd ed). 2008.

7.3.1 非饱和带水

在一些非均质环境中，可能存在上层滞水，即完全被非饱和带包围的饱和带，如图 7-6 所示。上层滞水的底部一般为不透水层。如图 7-6 所示，上层滞水在水位线之上，如果底部不透水层很宽，则上层滞水的储量足以让供水井在地下水位以上取水。

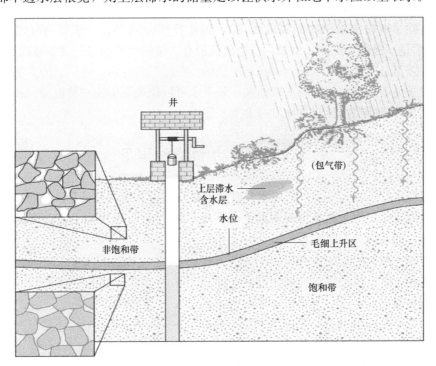

图 7-6 地下水的非饱和带、饱和带及上层滞水的剖面图

来源：DAVIS M L，MASTEN S J. Principles of Environmental Engineering and Science(3rd ed). 2014.

非饱和带的大部分水都存在于土壤中。土壤本质上是固体颗粒（矿物、有机物）、水和空气的统一体。考察土壤中水分首先要确定在一定时间间隔内有多少水进入土壤，即渗透速率。水进入土壤的速率取决于当前土壤的含水量和土壤传输水的能力。土壤含水量通常表示为土壤体积含水率或含水比（式 7-3）：

$$\theta = \frac{V_w}{V_t} \tag{7-3}$$

式中　θ——土壤的体积含水率（m^3/m^3）；

　　　V_w——土壤中水的体积；

　　　V_t——土壤的总体积。

土壤含水量也可用土壤重量含水率或饱和百分比来描述。土壤重量含水率是土壤中水分的重量与土壤总重量的比率。饱和含水量是土壤所能容纳的最大含水量。以饱和百分比表示的土壤含水量是判断土壤实际湿度的有效方法。

孔隙率是另一个重要的土壤性质。它是孔隙体积占土壤总体积的比例（式 7-1）。理论上，水可以填满土壤中的所有孔隙，因此，孔隙率为最大的潜在体积含水量。实际上，土壤的体积含水量很少能达到孔隙率值。水受重力作用迅速下降，土壤体积含水量也会迅速下降到孔隙率值以下。

7.3.2　饱和带水

非饱和带的水通过渗透下降至地下水位，就成了饱和带的水。饱和带的水移动缓慢，不能蒸发（除了通过深根植物的蒸腾作用），因此停留时间很长。这就为存储更多洪积（即更大降雨量）提供了可能。例如，沙特阿拉伯有大量的"矿物水"储备可以利用。

然而，并不是所有地下水的运动都是缓慢的。在湿润的月份，地下水的大量流动和定期补充是很常见的。例如，在石灰岩地区，地下水可以像地下河流一样流动。在地表水文方面，地下水在维持夏季径流方面发挥着重要作用。

7.4　表面张力和毛细作用

因为极性水分子相互吸引，所以水具有的内聚力能将其结合在一起。水滴在空气中下落时容易形成球体，而水在平坦的表面上会形成水珠。在气-水界面上，表面似乎有一层薄膜被拉伸，因此称为表面张力。

一些表面上形成的水珠比其他表面上的水珠大得多。例如，新上蜡的汽车表面比没有上蜡的汽车表面的水珠更大。这是因为水分子对蜡中的非极性分子几乎没有吸引力。相比之下，水的自我吸引力相当大，这就使得蜡质表面上的水形成一个个的水珠。相反，如果把水滴在岩石表面上，它会扩散并润湿表面，不会冒出很多水珠。在含有空气和水的孔隙中，水通常会浸润在矿物表面，空隙的中心部分是空气，如图 7-7 所示。

水分子受到矿物表面的吸引力并在其表面拉

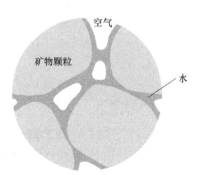

图 7-7　颗粒介质孔隙中的水和空气

来源：FITTS C R. Groundwater Science. 2002.

扯和扩散。一层大约厚 0.1~0.5μm 的水分子会被强烈吸引到介质表面，此部分水基本上是不运动的。距离介质表面约 0.5μm 的地方，引力将不足以阻止水分子的运动，在该距离之外的水分子可以自由移动。由于黏土矿物表面的带电性质，其对水分子的吸引力比其他矿物强。

由于水的表面张力作用，水内的压力小于孔隙内的气压。随着孔隙中水量的减少，矿物表面吸引力增大，水内压力减小，气-水界面形成与矿物颗粒相一致的扭曲形状。在部分饱和介质中，水对矿物表面的吸引力称为毛细作用力。毛细作用可使水浸润地下水位以上的孔隙。细颗粒材料中的毛细作用力往往较大，是因为其表面积大。

7.5 地下水运动

7.5.1 孔隙中的水传输

土壤的输水能力取决于其内部孔隙大小和孔隙之间的连通性。结构良好的土壤由稳定的团聚体组成，团聚体内部和团聚体之间存在大量不同尺寸的孔隙。在这种情况下，大孔隙至少占土壤体积的 10%。

这种结构为水流提供了相互连接的通道，且流速各不相同。生物的活动（如根和蠕虫）可以产生大孔隙，为水提供流动路径。大孔隙的类型有两种：一种是土壤基质中的大孔隙；另一种是从基质中分离出来的大孔隙。

（1）入渗

入渗是水进入土壤的净流量。当降雨速率超过入渗速率时，水向表层土壤的迁移速率通常随时间降低，直到达到一恒定值。入渗速率受降雨强度、土壤类型、地表条件和植被覆盖的影响。土壤孔隙中充满水和毛细

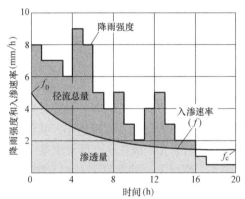

图 7-8 典型渗透曲线——渗透速率随时间的变化

血管作用的减弱导致了速率的暂时性下降。渗透速率随时间的变化如图 7-8 所示。当降雨速率超过入渗速率时，常用霍顿方程描述入渗速率（式 7-4）。

$$f = f_c + (f_0 - f_c)e^{-kt} \tag{7-4}$$

式中 f——入渗速率（或渗透能力）；

f_c——平衡、临界或最终入渗速率；

f_0——初始入渗速率；

k——经验常数；

t——时间。

该方程与许多其他质量平衡方程类似，即速率是指单位时间的距离，而不是单位时间的质量或体积。

如上所述，土壤类型影响水的入渗速率。因此，土壤越沙化，入渗速率就越大。土壤

压实度越高,或土壤黏粒含量越大,入渗速率越小。表 7-1 提供了一些常见土壤类型的参数。霍顿方程可用来估算给定的时间段内渗入水的总体积,即霍顿方程的积分形式(式 7-5):

$$V = A_s \int_0^t f \mathrm{d}t = A_s \int_0^t f_c + (f_0 - f_c)e^{-kt} \mathrm{d}t = A_s \left[f_c t + \frac{f_0 - f_c}{t}(1 - e^{-kt}) \right] \quad (7-5)$$

尽管霍顿方程适用于大多数土壤,但也存在一些局限性。对于沙质土壤,初始入渗速率 f_0 超过了大多数降雨强度。因此在这些情况下,霍顿方程会低估入渗速率。如前所述,渗透能力 f 随着孔隙的填充而减小,而不是随着时间的推移而减小;而在霍顿方程中,渗透能力是时间的函数,不是累积渗透量的函数。

一些典型土壤的霍顿方程参数   表 7-1

土壤类型	f_c(cm/h)	f_0(cm/h)	k(1/h)
Alphalpha 壤质砂土	3.56	48.26	38.29
Carnegie 砂壤土	4.5	35.52	19.64
Dothan 壤质砂土	6.68	8.81	1.4
Fuquay 含砾石壤质砂土	6.15	15.85	4.7
Leefield 壤质砂土	4.39	28.8	7.7
Troop 砂	4.57	58.45	32.71

(2)导水率

孔隙小的岩石或土壤只允许水缓慢迁移,而孔隙大、收缩性小的材料允许水分更快迁移。水在经过狭窄的小孔隙时,受到的剪切力大于水在经过较大孔隙时的剪切力。水的剪切力越大,黏性阻力越大,流速越小。其他因素不变时,地下水平均迁移速度与导水率(又称水力传导系数,K)成正比。导水率是一个经验常数,通常在实验室或现场试验中测得。

表 7-2 列出了一些常见岩石和土壤等典型介质的导水率数值范围。常见地质材料的导水率范围很大,最大相差 16 个数量级。

典型介质的导水率范围   表 7-2

材料	K(cm/s)
砾石	10^{-1} to 100
净砂	10^{-4} to 1
泥砂土	10^{-5} to 10^{-1}
泥沙	10^{-7} to 10^{-3}
冰堆物	10^{-10} to 10^{-4}
黏土	10^{-10} to 10^{-6}
石灰石和白云石	10^{-7} to 1
碎裂玄武岩	10^{-5} to 1
砂岩	10^{-8} to 10^{-3}
火成岩和变质岩	10^{-11} to 10^{-2}
页岩	10^{-14} to 10^{-8}

结晶岩中裂缝宽度和频率的巨大不同导致了导水率的范围如此之大。在碳酸盐岩被溶蚀侵蚀的地方，裂缝变宽形成大的开口，可形成"地下河"。由于熔岩流底部和顶部的开放型连续柱状节理和空隙，一些玄武岩也高度导水。玄武岩和石灰岩溶隙中的地下水流速比在典型介质的孔隙中的流速要高得多。

7.5.2 地下水的流动规律

在对地下水的研究中，一般会涉及地下水的迁移量和流速的问题。地下水污染修复方面的典型问题是："如何全部回收油井泄露的原油？"或者"受污染的地下水到达附近的河流需要多长时间？"

回答这些问题需进行地下水的流动分析，而地下水的流动分析基于控制地下水流的一些简单规律。达西定律和质量守恒是进行人工计算、计算机模拟与分析地下水流动过程的基础。

1856年，法国工程师亨利·达西负责法国第戎市的一个涉及使用砂粒过滤供水的项目。他在实验室进行了地下水运动的模拟实验，找到了控制水流通过砂粒速率的因素。此实验发现了地下水流动的基本原理，即广为所知的达西定律。

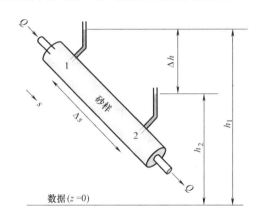

图 7-9 通过砂样的稳定流的示意图

注：压力计测量柱内位置 1 和 2 处的水头为 h_1 和 h_2。s 坐标方向与柱平行。

来源：FITTS C R. Groundwater Science. 2002.

达西的实验装置由一个填砂柱组成，柱上有一个入口和一个出口，与图 7-9 所示的结构类似。两个压力计（非常小的压力计）测量柱内两个点（h_1 和 h_2）的液压头。样本是饱和的，水流以流量为 Q 的速率稳定通过。

达西通过对某特定砂的反复试验发现，流量 Q 与两个压力计之间的压头差 Δh 成正比，与压力计之间的距离 Δs 成反比：

$$Q \propto \Delta h, Q \propto \frac{1}{\Delta s} \tag{7-6}$$

显然，流量 Q 也与砂柱的横截面积 A 成正比。结合这些观测结果，并以微分形式得出一维流动的达西定律：

$$Q_s = -K_s \frac{dh}{ds} A \tag{7-7}$$

式中 Q_s——含水层 s 方向的流量；

K_s——s 方向的导水率；

$\mathrm{d}h/\mathrm{d}s$——水力梯度；

A——横截面积。

式（7-7）中等式右边的减号是因为水头沿着水流方向降低。若水流沿 s 的正方向流动，则 Q_s 为正，$\mathrm{d}h/\mathrm{d}s$ 为负；若水流沿着与 s 相反的方向流动，则 Q_s 为负，$\mathrm{d}h/\mathrm{d}s$ 为正。水力坡降中的 h 项包括高程和水头。水头 h 和坐标 s 都是长度单位，因此 $\mathrm{d}h/\mathrm{d}s$ 是无量纲的。无量纲量 $\mathrm{d}h/\mathrm{d}s$ 表示水头沿 s 方向变化的速率，称为水力梯度。在非承压含水层中，水力梯度等于一定水平距离（即高程水头）上的地下水位下降的高度。在承压含水层中，水力梯度是潜水面（水平距离上）的落差。

7.5.3 含水层的补给和排泄

通常情况下，地下水是一直流动的，地下含水层通过补给和排泄与地表水发生作用。地下水和河流之间的作用相当复杂，且在很大程度上取决于当地的实际条件。水在自然条件下有汇入更快流速区域的趋势，因此，在河流枯水期，附近的地下水可以向上补给河流。但在河流丰水期，河流会通过渗透向下补给地下水。图 7-10 为地下水与河流相互作用的两种不同情况。图 7-10（a）中，由于地下水位较高，地下水补给河流；图 7-10（b）中，地下水位较低，河流补给地下水。主要河流的源头通常是远处的山脉，河流通常流经地下水位远低于河流水位的冲积平原。

饱和带地下水通常在低洼地区（泉水或地表水的底部）补给地表水。由于地下水总是朝着水头较低的方向移动，因此，此处地下水的水头高于补给口处的水头。

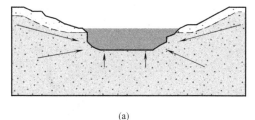

图 7-10 河流与地下水之间的相互作用
（a）地下水补给河流；（b）河流补给地下水
Source：DAVIE T. Fundamentals of Hydrology（2nd ed）. 2008.

图 7-11 展示了在一个假定的垂直剖面中地下水是如何从补给区流向排泄区的。补给水首先进入高地处的地下水系统，然后向地势低处（通常是湿地、泉水、溪流和湖泊）的排泄区迁移。

由于地质的异质性，水在地下的传播路径往往是不规则的。在具有高导水率的含水层中，水倾向平行于层边界流动。但在低导水率的含水层中，水倾向于以最短路径穿过该层，即几乎垂直于层边界流动。

在地下的浅层处，有许多小的局部流型，如图 7-11 所示。越往深处，地下水往往以大的区域模式流动，这反映了更大规模的地质和地形。在图 7-11 中，更深的水流集中在剖面右侧低洼的河流和洪泛平原。

地下水的停留时间（渗入后经过的时间）可从几天到几十万年不等。地下水在较大的

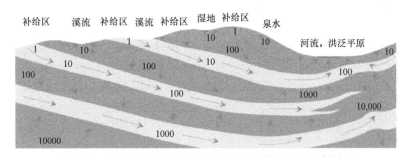

图 7-11 某假设环境中地下水补给区和排泄区的垂直剖面

注：箭头表示流向，数字表示地下水的停留时间（年）。浅色阴影表示含水层，深色阴影表示隔水层。

来源：FITTS C R. Groundwater Science. 2002.

盆地和水流系统较深部分的停留时间较长，在补给区和排泄区之间有较长的水流路径，流速较慢。如图 7-11 所示，含水层中的水比隔水层（半隔水层）中的水移动得更快。

大量的数学模型验证了地下的水位形态和非均质性对区域地下水流型的影响。模型为垂直面内的二维定常流。两个模型的结果如图 7-12 所示，地下水位上下起伏，在浅层形成局部流型；地下是不均匀的，浅色阴影区的渗透性是深色阴影区的 100 倍。这表明局部流动系统发生在浅层，非均匀性会对流型产生巨大影响。

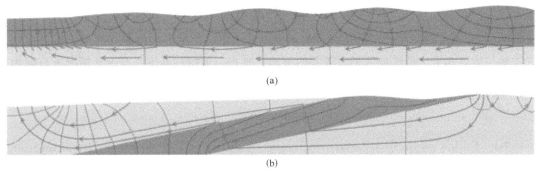

图 7-12 垂直面内稳定流动的两个数学模型

来源：FREEZE R A，WITHERSPOON P. Theoretical analysis of regional groundwater flow: Effect of water-table configuration and subsurface permeability variation. Water Resources Research，1967，3（2）：623-634.

泉水是地下水在地表的排泄地。在泉水处，地下水位与地表相交。泉水通常出现在陡坡底部、裂缝或含水层底部与斜坡相交的地方。图 7-13 显示了几种泉水的构造。

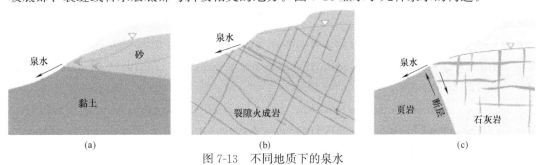

图 7-13 不同地质下的泉水

来源：FITTS C R. Groundwater Science. 2002.

在一些地区，地下水位在地表以下较深处，仅在主要河道与地表相交。这种情况发生在补给率相比于近地表层导水率较低时（图 7-14a）。地下水位较低且坡度较缓，从补给区到主要河流有较长的水流路径。在这种情况下，河流和湿地的距离会很大。

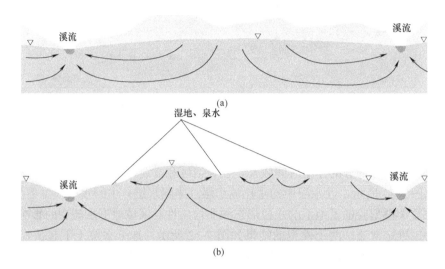

图 7-14　不同补给率和导水率所对应的地下水位位置
来源：FITTS C R. Groundwater Science. 2002.

图 7-14（b）显示了相反的情况，补水率比导水率高。地下水位要高得多，在许多地形洼地与地表相交。低洼地区形成潮湿的湿地、泉水、池塘和溪流。在这种情况下，会形成许多相对较短的局部流动路径。

Questions

7-1　Define the term saturated hydraulic conductivity and explain its importance in understanding groundwater flow.

7-2　Explain the terms confined and unconfined with respect to aquifers and describe how artesian wells come about.

7-3　Calculate the soil water content, given a soil sample $59.7\mathrm{m}^3$ in total volume, with 8.95kg water in it.

7-4　Determine the soil porosity, given the pores of the soil $5.54\mathrm{cm}^3$, with the total volume of the soil $15.98\mathrm{cm}^3$.

7-5　Consider Darcy's experiment shown in Figure 7-9. $h_1=95.0\mathrm{cm}$, $h_2=37.0\mathrm{cm}$, $\Delta s=20.0\mathrm{cm}$, $Q=3.5\mathrm{cm}^3/\mathrm{min}$, and the tube has a radius of 1.5cm. Calculate K_s and give your result in cm/s.

7-6　Determine the discharge of flow through a well-sorted grave aquifer, given that $K=0.01\mathrm{cm/s}$. The change in head is 1m over a distance of 1500m, and the cross-sectional area of the aquifer is $750\mathrm{m}^2$.

7-7　Define the following terms: aquifer, vadose zone, unsaturated zone, zone of aeration, phreatic zone, aquitard, aquiclude, perched aquifer, confined aquifer, unconfined aquifer, hydraulic conductivity, hydraulic gradient, porosity, piezometric head (surface).

7-8 A gravelly sand has hydraulic conductivity of 6.1×10^{-4} m/s, a hydraulic gradient of 0.00141 m/m and a porosity of 20%. Determine the Darcy velocity and the average linear velocity.

7-9 What factors will affect infiltration rates?

Chapter 8 WATER RESOURCES MANAGEMENT IN A CHANGING WORLD

We live in a world that is constantly on the change. This applies from the natural, through to the economic world and is fundamental to the way that we live our lives. The theory of evolution proposes that in order to survive each species on the planet is changing over a long time period (through natural selection) in order to adapt to its ecosystem fully. Equally, economists would say that people and businesses need to adapt and change to stay competitive in a global economy. If water were fundamental to all elements of our life on this planet, then we would expect to see that hydrology constantly is on the move to keep up with our changing world. It is perhaps no great surprise to say that hydrology has been, and is changing——but not in all areas. The principle of uniformitarianism states in its most elegant form: "the present is the key to the past". Equally, it could be said that the present is the key to the future and we can recognize this with respect to the fundamentals of hydrology. After a few decades, people may be living in a different climate from now, their economic lives may be unlike ours, and almost certainly, their knowledge of hydrological processes will be greater than ours. However, the hydrological processes will still be operating in the same manner, although may be at differing rates than those that we measure today. The early chapters of this book have been concerned with hydrological processes and our assessment of them. In this chapter, several hydrological issues are explored with respect to managing water resources and change that might be expected.

The issue of change is explored in a water resource management context: how we respond to changes in patterns of consumption; increasing population pressure and possible changes in climate. The topics discussed here are not exhaustive in covering all issues of change that might be expected in the near future, but they do reflect some of the major concerns.

8.1 Hydrology and Change

In the field of water resource management, there is a problem concerning the statistical techniques that we use. In a frequency analysis technique there is an inherent assumption that a storm event with similar antecedent conditions, at any time in the streamflow record, will cause the same size of storm. We assume that the hydrological regime is stationary with time.

Under conditions of land use or climate change, it is quite possible that these conditions will not be met. In China, for example, large-scale urbanization in recent years (skyscrapers, viaducts, impermeable surfaces, etc.) has led to changes in land use. This makes it difficult to put much faith in hydrology techniques such as frequency analysis when it is known that the hydrological regime has changed during the period of record. These are the types of challenges facing water resource management in an ever-changing world. The following section outlines some of the changes possible and uses case studies to demonstrate the possible effects of those changes.

8.1.1 Climate Change

At the start of the 21st century, climate change is the biggest environmental challenge, dominating the scientific media and research agenda. Humans link any unusual weather patterns to the greenhouse effect and its enhancement. The summer of 2006 in Northern Europe was one of the hottest and driest on record and there was drought. At the same time, New Zealand experienced one of the wettest winters on record with record snowfalls to sea level, followed by a wet and cold summer. At various times in the media, both these events were linked to global warming.

The difficulty with trying to verify any real link to climate change is that hydrological systems naturally contain a huge amount of variability. The extreme events we are experiencing now may be part of that natural variability, or they may be pushed to further extremes by climate change. It is unlikely that we will know for sure until it is too late to try to do anything about it.

Predictions from the Intergovernmental Panel on Climate Change (IPCC, 2007) suggest that the Earth may experience a global surface temperature rise of $0.2^\circ C$ per decade over the next 100 years. Even if the concentrations of all greenhouse gases and aerosols had been kept constant at the year 2000 levels, a further warming of about $0.1^\circ C$ per decade would be expected. Linked to this prediction are an increase in sea level of $15 \sim 95$cm and changes in the temporal and spatial patterns of precipitation.

All of these predicted changes will influence the hydrological cycle in some way, but it is difficult to pinpoint exactly how. At the very simple level, a temperature rise would lead to greater evaporation rates, which in turn puts more water into the atmosphere. This may lead to changes in precipitation patterns. How this affects the hydrology of an individual river watershed is very difficult to predict. The most common method to make predictions is to take the broad-brush predictions from a global circulation model (often at scale of 1° latitude and longitude per grid square) and downscale it to the local river watershed level.

Currently, it is difficult to make specific predictions for changes in hydrology, as the feedback mechanisms within climate change are not properly understood.

8.1.2 Change in Land Use

The implications of land use change for hydrology has been an area of intense interest to research hydrologists over the last fifty or more years. Issues of land use change affecting hydrology include increasing urbanization, changing vegetation cover, land drainage and changing agricultural practices leading to salination.

(1) Vegetation Change

Trees have a heavy effect on evaporation and interception rates. This is a hydrological impact of vegetation cover change. In general, the greater the amount of deforestation is, the larger the subsequent streamflows will be, but the actual amount is dependent on the vegetation type and precipitation amount. While it may be possible to say that in general a land use change that has increased tree cover will lead to a water loss, it is not easy to predict by how much that will be.

Some studies show that with the loss of forest cover both low flows and peak flows increase. The low flow response is altered primarily through the increase in water infiltrating to groundwater without interception by a forest canopy. The peak flow response is a result of a generally wetter soil and a low interception loss during a storm when there is no forest canopy cover. The time to peak flow may also be affected, with a more sluggish response in a watershed with trees. In a modeling study, it was suggested that any changes in peak flow that result from afforestation are not gradual but highly dependent on the timing of canopy closure.

(2) Land Drainage

Land drainage is a common agricultural "improvement" technique in areas of high rainfall and poor natural drainage. In an area such as the Fens of Cambridgeshire, Norfolk and Lincolnshire in England, this has taken the form of drains or canals and an elaborate pumping system, so that the natural wetlands have been drained completely. The result of this has been the utilization of the area for intensive agricultural production since the drainage took place in the seventeenth and eighteenth centuries. Since that time, the land has sunk, due to the removal of water from the peat-based soils, and the area is totally dependent on the pumping network for flood protection. To maintain this network vegetation control and clearance of silt within channels is required a cost that can be challenged in terms of the overall benefit to the community.

At the smaller scale, land drainage may be undertaken by farmers to improve the drainage of soils. This is a common practice throughout temperate regions and allows soils to remain relatively dry during the winter and early spring. The most common method of achieving this is through a series of tile drains laid across a field that drain directly into a watercourse (often a ditch). Traditionally tile drains were clay pipes that allowed water to drain into them through the strong hydraulic gradient created by their easy drainage towards the ditch.

Modern tile drains are plastic pipes with many small holes to allow water into them. Tile drains are normally laid at about 60cm in depth and should last for at least fifty years or more. To complement the tile drains "mole drainage" is carried out. This involves dragging a large, torpedo-shaped metal "mole" behind a tractor in lines orthogonal to the tile drains. This creates hydrological pathways, at 40～50cm in depth, towards the tile drains. Mole draining may be a regular agricultural activity, sometimes every two to five years in heavy agricultural land (i.e. clay soils). Normal plough depth is around 30cm, so that the effects of mole draining last beyond a single season.

The aim of tile and mole drainage is to hold less water in a soil. This may have two effects on the overall hydrology. It allows rapid drainage from the field, therefore increasing the flashy response (i.e. rapid rise and fall of hydrograph limbs) in a river.

A large plough creates drains in an area, with the seedlings being planted on top of the soil displaced by the plough (i.e. immediately adjacent to the drain but raised above the water table). Like all land drainage, this will lower the water table and allow rapid routing of stormflow.

(3) Salination
Salination is an agricultural production problem that results from a buildup of salt compounds in the surface soil. Water flowing down a river is almost never "pure"; it will contain dissolved solids in the form of salt compounds. These salt compounds are derived from natural sources such as the weathering of surface minerals and sea spray contained in rainfall. When water evaporates, the salts are left behind, something we are familiar with from salt lakes such as in Utah, central Australia, and the Dead Sea in the Middle East. The same process leads to salinity in the oceans.

Salination of soils (often also referred to as salinization) occurs when there is an excess of salt rich water that can be evaporated from a soil. The classic situation for this is where river-fed irrigation water is used to boost agricultural production in a hot, dry climate. The evapotranspiration of salt rich irrigation water leads to salt compounds accumulating in the soil, which in turn may lead to a loss of agricultural production as many plants fail to thrive in a salt-rich environment. Although salination is fundamentally an agronomic problem, it is driven by hydrological factors (e.g. water quality and evaporation rates), hence the inclusion in a hydrological textbook.

8.1.3 Groundwater Depletion

In many parts of the world, there is heavy reliance on aquifers for provision of water to a population. In England, around 30% of reticulated water comes from groundwater, but that rises to closer to 75% in parts of southeast England. The water is extracted from a chalk aquifer that mostly receives a significant recharge during the winter months. Apart from very dry periods, there is normally enough recharge to sustain withdrawals. Not all groundwater is recharged so readily. Many aquifers have built up their water reserves over millions of years and receive very little infiltrating rainfall on a year-by-year basis. Much of the Arabian Peninsula in the Middle East is underlain by such an aquifer. The use of this water at high rates may lead to groundwater depletion, a serious long-term problem for water management.

8.1.4 Urbanization

Many aspects of urban hydrology have already been covered, especially with respect to water quality, but the continuing rise in urban population around the world makes it an important issue to consider under the title of change. There is no question that urban expansion has a significant effect on the hydrology of any river draining the area. Initially this may be due to climate alterations affecting parts of the hydrological cycle. The most obvious hydrological impact is on the runoff hydrology, but other areas where urbanization may have an impact are point source and diffuse pollution affecting water quality, river channelization to control flooding, increased snow melt from urban areas and river flow changes from sewage treatment.

(1) Urban Climate Change

In Table 8-1, some of the climatic changes due to urbanization are expressed as a ratio between the urban and rural environments. This suggests that within a city there is a 15% reduction in the amount of solar radiation reaching a horizontal surface, a factor that will influence the evaporation rate. Studies have also found that the precipitation levels in an urban environment are higher by as much as 10%. Atkinson (1979) detected an increase in summer thunderstorms over London, which was attributed to extra convection and condensation nuclei being available. Other factors greatly affected by urbanization are winter fog (doubled) and winter ultraviolet radiation (reduced by 30%).

Chapter 8 WATER RESOURCES MANAGEMENT IN A CHANGING WORLD

Difference in Climatic Variables between Urban and Rural Environments   Table 8-1

Climatic Variable	Ratio of City to Environs
Solar radiation on horizontal surfaces	0.85
UV radiation: summer	0.95
UV radiation: winter	0.70
Annual mean relative humidity	0.94
Annual mean wind speed	0.75
Speed of extreme wind gusts	0.85
Frequency of calms	1.15
Frequency and amount of cloudiness	1.10
Frequency of fog: summer	1.30
Frequency of fog: winter	2.00
Annual precipitation	1.10
Days with less than 5mm precipitation	1.10

(2) Urban Runoff Change

The changes in climate are relatively minor compared to the impact that impermeable surfaces in the urban environment have on runoff hydrology. Roofs, pavement, roads, parking lots and other impermeable surfaces have extremely low infiltration characteristics, consequently Hortonian overland flow readily occurs. These surfaces are frequently linked to gutters and stormwater drains to remove the runoff rapidly. The result of this is far greater runoff and the time to peak discharge being reduced.

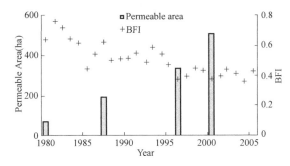

Figure 8-1 Baseflow Index (BFI——Proportion of Annual Streamflow as Baseflow) with Time in a Small Watershed in Auckland, New Zealand where There has been Steady Urbanization
Note: The vertical bars show area of permeable surfaces estimated from aerial photographs at 4 times.
Source: data courtesy of Auckland Regional Council.

Figure 8-1 shows some data from a steadily urbanizing watershed (13km^2) in Auckland, New Zealand. There has been a drop in the percentage of baseflow leaving the watershed (the baseflow index, BFI) which could be attributed to declining infiltration to groundwater and therefore less water released during the low flow periods. Care needs to be taken in interpreting a diagram like Figure 8-1 because it is also possible that the decline in BFI was caused by an increase in the stormflow and total flow and the actual amount of baseflow has stayed the same. Whichever way, there has been a change in the hydrological regime that can probably be attributed to the rise in permeable surfaces because of urbanization.

(3) Pollution from Urban Runoff

There is a huge amount of research and literature on the impacts of urbanization on urban water quality. Some studies link the accumulation of heavy metals in river sediments to urban runoff, particularly from roads. Specific sources are tyre wear and vehicle brakes for zinc and buildings for lead, copper, cadmium and zinc.

The nature of urban runoff (low infiltration and rapid movement of water) concentrates the pollutants in the first flush of water. Studies have shown that over 80% of pollutant particles are washed into a drainage system within the first 6~10mm of rain falling, and often from a very small collection area within the urban watershed. This information is important when proposing strategies to deal with the urban pollutant runoff. One of the main methods is to create an artificial wetland within an urban setting so that the initial flush of storm runoff is collected, slowed down, and pollutants can be modified by biological action.

(4) River Channelization

It is a common practice to channelize rivers as they pass through urban areas in an attempt to lessen floods in the urban environment. Frequently, although not always, this will involve straightening a river reach and this has impacts on the streamflow. The impact of urban channelization is not restricted to the channelized zone itself. The rapid movement of water through a channelized reach will increase the velocity, and may increase the magnitude of a flood wave travelling downstream. Deposition of sediment downstream from the channelized section may leave the area prone to flooding through a raised riverbed.

(5) Urban Snow Melt

The influence of urbanization on snow melt is complicated. Melt intensities are generally increased in an urban area, although shading may reduce melt in some areas. Overall, there is a greater volume of water in the early thaw from an urban area when compared to a rural area. This may be complicated by snow clearance operations, particularly if the cleared snow is placed in storage areas for later melting. In this case, the greater mass of snow in a small area will cause a slower melt than if it were distributed throughout the streets.

(6) Waste Water Input and Water Extraction

Human intervention in the hydrological regime of a river may be in the form of extraction (for irrigation or potable supply) or additional water from waste water treatment plants. The amount of water discharged from a sewage treatment work into a river may cause a significant alteration to the flow regime. At periods of low flow rivers may comprise water derived from waste water effluent. The extra flow that a river derives from sewage effluent may be especially significant if the waste water effluent has been abstracted from another watershed.

8.2 Water Resource Management

Water is a resource because we need it, and there are ways that we can manipulate its provision, therefore, water resource management is a very real proposition. We have an obligation to protect the water resource for future generations and for other species that coexist with the water. Therefore, water resource management needs to embrace sustainable development in its good practices.

Almost all of the processes found in the hydrological cycle can be manipulated in some way, and any types of human intervention may have a significant impact on water resources. Table 8-2 sets out some of these interventions and the implications of their being dealt with by those involved in water resource management. It is immediately apparent from Table 8-2 that the issues go far beyond the river boundary. For example, land use change alters infiltration rates, which control water storage and runoff generation of local areas. Therefore, any decisions on land use need to include consultation with water resource managers.

A key part of water resource management involves water allocation: the amount of water made available to users, including both out of stream use (e.g. irrigation, town water supply) and instream environmental use (e.g. amenity values, supporting aquatic populations). Water allocation in a resource management context is about how to ensure fair and equitable distribution of the water resource between groups of stakeholders.

Manipulation of Hydrological Processes of Concern to Water Resource Management

Table 8-2

Hydrological Process	Human Intervention	Impact
Precipitation	Cloud seeding	Increase rainfall
Evaporation	Irrigation	Increase evaporation rates
	Change vegetation cover	Alter transpiration and interception rates
	Change rural to urban	Increase evaporation rates

Continued

Hydrological Process	Human Intervention	Impact
Storage	Change land use	Alter infiltration rates
	Aquifer storage and recovery (ASR)	Manipulate groundwater storage
	Land drainage	Lower local water tables
	Building reservoirs	Increase storage
Runoff	Change land use	Alter overland flow rates
	Land drainage	Rapid runoff
	River transfer schemes	Alter river flow rates
	Water abstraction	Remove river water and groundwater for human consumption

Another key part of water resource management is the involvement of many different sectors of the community in decision-making. This has led to different approaches to water management that stresses integration between different sectors. There are three key concepts in this area: Integrated Water Resource Management (IWRM), Integrated Catchment Management (ICM) and Stormwater Management, which are discussed in more detail below.

8.2.1 Integrated Water Resource Management (IWRM)

The concepts behind IWRM lie in the so-called "Dublin Principles". In January 1992, 500 participants, including government-designated experts from 100 countries and representatives of 80 international, intergovernmental and non-governmental organizations attended the International Conference on Water and the Environment in Dublin, Ireland.

The conference adopted what has been termed "the Dublin Statement", which was taken forward to the Earth Summit Conference in Rio de Janiero, Brazil later that year. The Dublin Statement established four guiding principles for managing freshwater resources, namely:

(1) Fresh water is a finite and vulnerable resource, essential to sustain life, development and the environment.
(2) Water development and management should be based on a participatory approach, involving users, planners and policy makers at all levels.
(3) Women play a central part in the provision, management and safeguarding of water.
(4) Water has an economic value in all its competing uses and should be recognized as an

economic good.

These four principles underlie IWRM, especially the concepts of a participatory approach and that water has an economic value. An economic good, as used in principle four, is defined in economics as: a physical object or service that has value to people and can be sold for a non-negative price in the marketplace. A major implication from principle four is that water is not a gift or a free right to any water user, it needs to be recognized that using water restricts the usage by others and therefore there is a cost involved in the action.

Eight IWRM Instruments for Change as Promoted by the Global Water Partnership

Table 8-3

IWRM Instrument for Change	Comments and Requirements
Water resources assessment	Understanding what water resources are available and the water needs of communities. Requiring measurements of flows, groundwater levels, etc. and water usage (e.g. metering of take)
IWRM plans	Combining development options, resource use and human interaction. Requiring inter-sectoral approach
Demand management	Using water more efficiently. Requiring knowledge of where water losses occur (leakage) and plans on how to promote water efficiency
Social change instruments	Encouraging a water-oriented society. Requiring community education on the importance of using water wisely
Conflict resolution	Managing disputes, ensuring sharing of water. Requiring promotion of trust between sectors and robust dispute settlement systems
Regulatory instruments	Allocation and water use limits. Requiring good knowledge about the amount of available resource and how the hydrological system responds to stress (either natural or human-induced)
Economic instruments	Using value and prices for efficiency and equity. Requiring good information on water usage and overall water demand
Information management and exchange	Improving knowledge for better water management. Requiring good data-sharing principles (e.g. between flood control and water supply agencies)

Source: Global Water Partnership. 2004. http://www.gwp.org/.

The emphasis within an IWRM approach to water management is on integration between sectors involved in water resources, including local communities (a participatory approach). Although this is promoted as a new approach to resource management, it is in many ways a return to traditional values with recognition of the interconnectedness of hydrology, ecology and land management. If there is a large amount of water from a stream allocated to agriculture, there is less available for town water supply and instream ecology. IWRM is a framework for change that recognizes this interconnectedness and builds structures to manage water with this in mind. It is an attempt to move away from structures that promote individual sectors competing against each other for the scarce resource of water and moves towards joint ownership of water resource management.

The types of approaches suggested for use within IWRM are illustrated in Table 8-3. These are eight instruments for change that the Global Water partnership promotes as being integral to IWRM. While these are by no means the only ways of achieving an integrated management of water resources, they are a useful starting point.

8.2.2 Integrated Catchment Management (ICM)

Integrated Catchment Management (also sometimes referred to as Integrated Water Basin Management, IWBM) is essentially a subset of IWRM. It aims to promote an integrated approach to water and land management but with two subtle differences:

(1) ICM recognizes the catchment (or river basin) as the appropriate organizing unit for understanding and managing water-related biophysical processes in a context that includes social, economic and political considerations;
(2) There is recognition of the spatial context of different management actions and in particular the importance of cumulative effect within a catchment.

By defining a river catchment as the appropriate organizing unit for managing biophysical processes, there is a recognition that hydrological pathways are important and these provide an appropriate management, as well as biophysical boundary. Cumulative effect refers to the way in which many small actions may individually have very little impact but when combined the impact may be large. This is true for a river catchment system where individual point discharges of pollution may be small but when accumulated within the river they may be enough to cross an environmental threshold.

The word "integrated" in an ICM context is defined using three different connotations:

(1) Integration between the local community, science and policy so that the community is linked into the planning and execution of both science and policy and scientific research is being carried out in an environment close-linked into policy requirements and vice versa (Figure 8-2);

(2) Integration between different scientific and technical disciplines to tackle multi-dimensional problems;

(3) Spatial integration throughout a watershed so that the cumulative impact of different actions can be assessed.

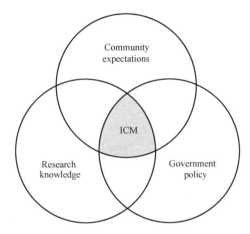

Figure 8-2   The Integrating Nature of ICM within the Context of Science,
Local Community and Governance
Source: DAVIE T. Fundamentals of Hydrology (2nd ed). 2008.

8.2.3 Stormwater Management

Stormwater discharges, which are considered point sources, were traditionally collected in piped networks and transported off site as quickly as possible, either directly to a stream or river, to a large stormwater retention or detention basin, or to a combined sewer system flowing to a wastewater treatment plant. Stormwater retention or detention basins detain and slow the flow of stormwater, allowing larger and heavier material to settle out and chemicals and smaller particles to be filtered out before the water is discharged into receiving waters. These ponds reduce the likelihood of flooding and reduce the effects of urbanization on water quality. However, these traditional stormwater infrastructures reduced the extent of pervious surfaces, and impaired infiltration and, therefore, groundwater recharge. They also consume wildlife habitat and available space for recreation or other needs. Figure 8-3 shows how urbanization, resulting in the rapid flow of stormwater into storm drains, sewer systems and drainage ditches, affects the stream flow hydrograph and causes flooding.

8.2.3.1 Sponge City

The goal of sponge city practices is to reduce runoff volume and enhance the filtration and removal of pollutant from stormwater. These practices include bioretention facilities or rain gardens, grass swales and channels, vegetated rooftops, rain barrels, cisterns,

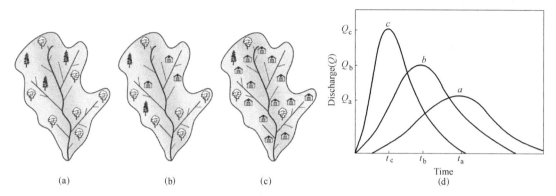

Figure 8-3　Effect of the Watershed on a Hydrograph
(a) Undeveloped; (b) Partially Developed; (c) Fully Developed; (d) $Q_c>Q_b>Q_a$ and $t_c<t_b<t_a$
Source: DAVIS M L, MASTEN S J. Principles of Environmental Engineering and Science (3rd ed). 2014.

vegetated filter strips and permeable pavements. At both the site and regional scale, sponge city practices aim to preserve, restore and create green space using soils, vegetation and rainwater harvest techniques. Sponge city is an approach to land development (or re-development) that works with nature to manage stormwater as close to its source as possible. Sponge city employs principles such as preserving and recreating natural landscape features, minimizing effective imperviousness to create functional and appealing site drainage that treat stormwater as a resource rather than a waste product. By implementing sponge city principles and practices, water can be managed in a way that reduces the impact of built areas and promotes the natural movement of water within an ecosystem or watershed. Applied on a broad scale, sponge city can maintain or restore a watershed's hydrologic and ecological functions.

Bioretention facilities typically contain these six components: grass buffer strips, sand bed, ponding area, organic layer, planting soil and vegetation. The grass buffer strips serve to reduce the velocity of the runoff and filter particulate matters from the water. The sand bed helps to aerate the water and drain the planting soil. It also helps to flush pollutants from soil materials. The ponding area provides storage of excess runoff, especially for the initial flush during a rainfall event. It also facilitates the settling of particulate material and the evaporation of water. The organic layer provides a support medium for microbiological growth, which decomposes organic material in the storm water. This layer also serves as a sorbent for heavy metals and other hydrophobic pollutants. The plants, which grow in the planting soil, take up nutrients and aid in the evapotranspiration of water. The soil provides additional water retention and may absorb some pollutants, including hydrocarbons and heavy metals.

Green roofs effectively reduce urban stormwater runoff by reducing the amount of impervious surfaces. They are especially effective in older urban areas, where the percentage of land that impervious to water flow is high. A green roof is constructed of multiple layers: a vegetative layer, media, a geotextile and a synthetic drain system. Green roofs also extend the life of the roof, reducing energy costs and conserving valuable land that would otherwise be required for stormwater runoff controls.

Permeable pavements can be used to reduce the percent of impervious surfaces in a watershed. Porous pavements are most appropriate for low traffic areas, such as parking lots and sidewalks. They have been successfully installed in coastal areas with sandy soils and flatter slopes. The infiltration of stormwater into underlying soils promotes pollutant treatment and groundwater recharge.

Other techniques used in sponge city, including grass swales and channels, rain barrels, cisterns, vegetated filter strips, serve to redirect runoff from sewers and stormwater collection systems. Rain barrels and cisterns allow for the collection of water and its subsequent use for lawn and garden irrigation and nonpotable water uses such as toilet flushing. Grass swales and channels, along with vegetated filter strips help reduce the amount of impervious surface in development areas and enhance infiltration and groundwater recharge.

8.2.3.2 Constructed Wetlands

As sponge city, constructed wetlands seek to maintain or restore the natural hydrology of the site during wet weather. It is designed to mimic natural wetlands to retain stormwater and remove pollutants by gravity settling, sorption, biodegradation and plant uptake.

The velocity of water slows down as it flows through a wetland. This allows suspended solids to either become trapped by vegetation and to settle by gravity. Hydrophobic organic pollutants and heavy metals can sorb to plant or soil organic matter. Biodegradable pollutants can be assimilated and transformed by plants or microorganisms. Nutrients are absorbed by wetland soils and taken up by plants and microorganisms. For example, microbes found in wetlands can convert organic nitrogen into useable, inorganic forms (i.e. NO_3^- and/or NH_4^+), which is essential for plant growth. Subsequent reactions (known as denitrification) can convert the nitrate to nitrogen, which can be safely released to the atmosphere. Phosphorus can be assimilated by microorganisms and incorporated into cellular biomass.

Constructed wetlands can also be a cost-effective and technically feasible approach to treating stormwater. Wetlands are often less expensive to build than traditional treatment options, which have lower operating and maintenance costs and can better handle fluctuating water levels.

Additionally, they are aesthetically pleasing and can promote water reuse, wildlife habitat and recreational use. Constructed wetlands should be built on uplands and outside floodplains or floodways in order to prevent damage to natural wetlands. Water control structures need to be installed to ensure the desired hydraulic flow patterns. Where the soils are highly permeable, an impervious, compacted clay liner should be installed to protect underlying groundwater. The original soil can be placed over the liner. Wetland vegetation is then planted or allowed to establish naturally.

However, there are limitations to constructed wetlands for stormwater management. Highly variable flow conditions may make it difficult to maintain vegetation, although the recycle of water through the wetland can help ameliorate problems associated with low flow conditions. The detained water may act as a heat sink, resulting in the discharge of significantly warmed water to downstream water bodies. In addition, until vegetation is established and during cold seasons the removal of pollutants is likely to be minimal. However, with careful design and proper maintenance, constructed wetlands can cost-effectively remove pollutants from stormwater for many years.

Summary

It has shown that there are many aspects of change in hydrology to be considered. Equally, there are different management structures and principles that can be used to manage the change. In order to understand and make predictions concerning change, it is essential to understand the fundamentals of hydrology: how processes operate in time and space; how to measure and estimate the rates of flux for those processes; and how to analyze the resultant data. One thing should be borne in mind that the fundamental process is constant, and it is their rates of flux in different locations that always change. It is fundamentally important for hydrology as a science to investigate these rates of change, and finding new ways of looking at the scales of change in future.

Translation of Some Sections

部分章节参考译文

8.2 水资源管理

水是一种资源，由于人类的需要，可以通过各种方式来调节它的供应，因此，水资源管理是一个非常现实的问题。我们有义务为子孙后代和与水共存的其他物种来保护水环境。因此，对水资源的管理应并入可持续发展。

水文循环中几乎所有环节都可以用某种方式加以人为干预，并且任何形式的人为干预都可能对水资源产生重大影响。表 8-2 列举了一些水资源管理干预的措施及其影响。例如，土地利用的变化会改变地面渗透率，从而影响当地的储水量和径流形成量，因此有关土地利用的决策应有水资源管理者的参与。

与水资源管理有关的水文过程干预　　　　　　　　表 8-2

水文过程	人为干预	影响
降水	云种散播	增加降雨
蒸发	灌溉	提高蒸发率
	改变植被覆盖类型	改变蒸腾和截留速率
	城镇化	提高蒸发率
储存	改变土地利用类型	下渗速率变化
	含水层储存和回采（ASR）	地下水存储量变化
	地面排水	地下水位降低
	修建水库	增加蓄水量
径流	改变土地利用类型	地表径流速率改变
	地面排水	径流速率提升
	河流运输	河流流速改变
	取水	地表及地下水转移

水资源管理的一个关键是进行水资源分配（即向用户提供用水服务），包括河道外使用（如灌溉、城镇供水）和河道内环境利用（如建立便利设施、支持渔民）。资源管理中的水资源分配涉及水资源在利益相关群体之间公平分配的问题。

水资源管理的另一个关键是允许不同的社区部门参与决策。不同部门之间的整合导致了不同的水资源管理方法。水资源管理领域有 3 种重要思路：水资源综合管理（IWRM）、流域综合管理（ICM）和雨水管理，下文将对此进行更详细的讨论。

8.2.1　水资源综合管理（IWRM）

水资源综合管理即所谓的"都柏林原则"。1992 年 1 月，包括来自 100 个国家的政府指定专家和 80 个国际、政府间和非政府组织的共 500 名代表出席了在爱尔兰的都柏林市举行的国际水与环境会议。

会议通过了"都柏林声明"，并于当年在巴西的里约热内卢市举行的地球首脑会议上提出了这一声明。"都柏林声明"确立了管理淡水资源的 4 项指导原则，即：

（1）淡水是一种有限和脆弱的资源，对维持生命、社会发展和环境保护至关重要；
（2）水资源开发和管理应以共享为基础，让各级用户、规划者和决策者参与；
（3）妇女在水的供应、管理和保障方面起着中心作用；
（4）水在所有竞争性用途中均具有经济价值，应被视为经济商品。

这 4 项原则是进行水资源综合管理的基础，尤其是共同参与的理念和水具有经济价值的理念。第 4 项原则中所提到的经济商品，在经济学中被定义为：对人们有价值并且可以在市场上以非负价格出售的实物或服务。其主要含义是：水不是给予任何用水者的礼物或免费的权利。人们需要认识到，自己用水就限制了其他人的使用，用水行为是需要付出代价。

水资源综合管理的重点是水资源各部门（包括地方社区）之间的一体化。虽然这是一种新的资源管理方法，但因其认识到水文、生态和土地管理之间的相互联系，可以说它在许多方面是传统价值观的回归。如果把一条河流中大多数的水分配给农业，那么城镇供水和河道内生态系统可用的水量就会减少。水资源综合管理是一个变革框架，它认识到了水与其他部门的相互联系，并在此基础上建立了管理水资源的结构。此框架在构建过程中，摒弃了那些促进各个部门相互竞争以争夺稀缺水资源的结构，吸纳了水资源共同管理的结构。

水资源综合管理中的建议措施如表 8-3 所示。这是促进全球水事伙伴关系的八项变革文书，是水资源综合管理的组成部分。这虽不是实现水资源综合管理的唯一途径，但是一个很好的起点。

全球水事伙伴关系推动的八项水资源综合管理变革文书 表 8-3

水资源综合管理变革手段	意见和要求
水资源评价	充分了解可利用的水资源和社区对水的需求；测量流量、地下水位等和用水情况（例如测量取水量）
水资源综合管理计划	综合发展选择、资源使用和人的相互作用；跨部门合作
需求管理	更有效地利用水资源； 了解哪里发生了水的损失（渗漏）以及提供提高用水效率的方案
社会变革手段	鼓励建设以水为导向的社会； 对社区进行明智用水的重要性教育
冲突解决	管理争端，确保水资源共享； 加强各部门之间的信任和建立健全的争端解决体系
监管手段	分配和限制用水； 充分了解可利用资源的数量，以及水文系统如何应对供水压力（自然或人为）
经济手段	使用价值和价格手段促进效率和公平； 掌握关于用水情况和总体用水需求的有效信息
信息管理与交换	提升管理知识水平； 建立良好的数据共享原则（例如在洪水控制和水供给机构之间）

8.2.2 流域综合管理（ICM）

流域综合管理本质上是水资源综合管理的一个子集，目的是推动对水和土地进行综合管理，但二者有两个细微的区别：

（1）流域综合管理认为流域是理解和管理与水相关的生物物理过程的适当的单元，包括社会、经济和政治考量；

（2）人们认识到不同管理行动的空间尺度，尤其是流域区内累积效应的重要性。

通过将流域定义为管理生物物理过程的单元，人们认识到水文路径的重要性，这些路径提供了合适的管理范围和生物物理学边界。累积效应是指单独作用时影响很小，但当组合起来时影响可能很大的效应。这也适用于单个点的污染排放量可能很小，但当其在河流中累积时可能超过环境阈值的流域系统。

流域综合管理中的"综合"一词有 3 种不同的含义：

（1）社区目标、科学研究、政府政策的 3 项整合，使社区规划和实际与科学理论和政策相关联，使科学研究在符合政策要求的环境中进行（图 8-2）；

（2）不同学科之间相互融合，以解决多维问题；

（3）整个流域进行空间整合，以评估不同行为的累积影响。

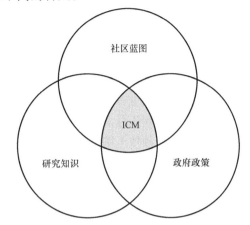

图 8-2 ICM 在科学、社区和政府间的综合性
来源：DAVIE T. Fundamentals of Hydrology（2nd ed）. 2008.

8.2.3 雨水管理

传统上，雨水通过管道网络收集，直接输送至河（溪）流、大型雨水蓄水池、滞留池，或流入污水处理厂的下水道系统。滞留池可减缓流速，并在接纳水体进入前使大且重的物质得到沉淀，小颗粒和化学物质得到过滤。滞留池降低了洪水的可能性，减少了城市化对水质的影响。然而，这些传统的雨水设施减少了透水表面的范围，不仅不利于地表渗透和地下水补给，还占用了野生动物生存生境、休息场所或其他活动需要的空间。图 8-3 通过对比某流域不同城镇化水平所对应的不同水文过程线反映出城镇化水平对洪水形成的影响。

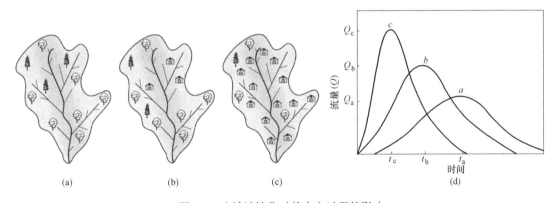

图 8-3 流域城镇化对其水文过程的影响
（a）未发展；（b）部分发展；（c）完全发展；（d）流量对比
来源：DAVIS M L, MASTEN S J. Principles of Environmental Engineering and Science（3rd ed）. 2014.

8.2.3.1 海绵城市

海绵城市的目标是减少径流量，加强雨水的下渗和其中污染物的过滤与去除。通常的做法是建设生物滞留设施或雨水花园、草沟和沟渠、植被屋顶、雨水桶、蓄水池、植被过滤带和透水路面。海绵城市旨在场地和区域范围内利用土壤、植被和雨水收集技术来保护、恢复和创造绿色空间，是土地开发（或再开发）的一种方法，它与自然一起尽可能在源头管理雨水。海绵城市基于保护和重现自然景观特征、最大限度地降低不渗透性等原则，创建具有功能性和景观性的排水系统，将雨水作为资源而不是废弃物。通过建设海绵城市，减少洪水对建成区的影响，促进生态系统或流域内水的自然流动，以维持或恢复流域的水文生态功能。

生物滞留设施通常由6个部分组成：草缓冲带、沙床、积水区、有机层、种植土壤和植被。草缓冲带起到降低径流速度和过滤水中颗粒物的作用。沙床可增加水中的空气量、排出种植土的水，以及冲刷土壤中的污染物。积水区可储存多余的径流，尤其是降雨过程中的初始冲刷。积水区也有助于颗粒物的沉降和水的蒸发。有机层为微生物的生长提供了介质支持，微生物有利于分解雨水中的有机物质。该层还可作为重金属和其他疏水性污染物的吸附剂。生长在种植土壤中的植物可吸收养分，促进水分蒸发。土壤提供额外的保水性，可吸收一些污染物，包括碳氢化合物和重金属。

绿色屋顶通过减少不透水表面，可以有效地减少城市雨水径流。一般老城区不透水的土地比例很高，因此绿色屋顶尤其有效。绿色屋顶由多层构成，包括：植被层、介质层、土工布层和合成排水系统。绿色屋顶还可以延长屋顶的使用寿命，降低能源成本，节约用地。

透水路面可用于降低流域中不透水地面面积的占比。多孔路面适合于低交通量区域，如停车场和人行道，在有沙土和较平坦斜坡的沿海地区有较多应用。雨水渗入下面的土壤可促进污染物的去除和地下水的补给。

其他海绵城市设施（包括植草沟和沟渠、雨水桶、蓄水池、植被过滤带）可用于下水道和雨水收集系统中径流的分流。雨水桶和蓄水池可以收集水，并随后用于草坪、花园灌溉以及其他非饮用水用途，如冲厕等。植草沟、沟渠和植被过滤带有助于减少不透水表面数量，增强渗透和地下水补给。

8.2.3.2 人工湿地

同海绵城市一样，人工湿地通过模拟自然湿地的重力沉降、吸附、生物降解和植物吸收来保留雨水和去除污染物，其旨在维护或恢复雨天的自然水文。

水流过人工湿地时流速减慢。这时悬浮固体要么被植被截留，要么在重力作用下沉降。疏水性有机污染物和重金属能吸附到植物或土壤有机体上。可生物降解污染物会被植物或微生物吸收和转化。养分会进入湿地土壤，并被植物和微生物吸收。例如，人工湿地中的微生物可以将有机氮转化为无机氮（即 NO_3^- 和 NH_4^+），这对植物生长至关重要。反硝化反应可以将硝酸盐转化为氮气，并释放到大气中。磷可以被微生物吸收。

人工湿地是一种成本低、效益高、技术可行的雨水处理方法。通常，人工湿地的建设、运行和维护成本比传统的处理方法低，而且能够更好地适应不断变化的水位。

人工湿地不仅增加了环境美感，而且可以促进水的再利用，增加野生动物栖息地和活动空间。人工湿地应建在高地、漫滩或洪水道以外，以防止其对自然湿地的破坏。同时，人工湿地需要安装水控制结构，以确保其运行所需的水力流动模式。如果土壤具有高渗透性，则应安装压实的不透水黏土衬垫，以保护地下水。原土可以放在衬垫上，然后种植湿地植被，或让其自然生长。

但是，人工湿地在雨水管理方面存在一定的局限性。尽管湿地的水循环有助于改善与低流量相关的问题，但高度变化的水流条件可能会难以维持植被生长。滞留的水可能会起到加热板的作用，使下游水体的水温升高。此外，在植被形成之前的寒冷季节，污染物的去除率是极小的。当然，经过精心的设计和适当的维护，人工湿地可以经济有效地去除雨水中的污染物。

Questions

8-1 Discuss how well the principles of Integrated Water Resource Management are applied to the management of a watershed near you.

8-2 Explain the way that human-induced climate change may affect the hydrological regime for a region.

8-3 Assess the role of land use change as a major variable in forcing change in the hydrological regime for a region near you.

8-4 Compare and contrast the impact of urbanization to the impact of land use change on general hydrology within the country where you live.

8-5 Discuss the major issues facing water resource managers over the next fifty years in a specified geographical region.

APPENDIX A: UNITS AND CONVERSIONS

Length Equivalents — Table A-1

Unit	Equivalent					
	mm	m	in.	ft	yd	mi
Millimeter	1	10^{-3}	0.0394	0.00328	0.00109	6.214×10^{-7}
Meter	10^3	1	39.37	3.281	1.0936	6.214×10^{-4}
Inch	25.4	0.0254	1	0.0833	0.02778	1.578×10^{-5}
Foot	304.8	0.3048	12	1	0.333	1.894×10^{-4}
Yard	914.4	0.9144	36	3	1	5.682×10^{-4}
Mile	1.609×10^6	1.609×10^3	6.336×10^4	5280	1760	1

Area Equivalents — Table A-2

Unit	Equivalent				
	in.^{2}	ft^2	m^2	acre	mi^2
Square inch	1	6.944×10^{-3}	6.452×10^{-4}	1.59×10^{-7}	2.491×10^{-10}
Square foot	144	1	0.0929	2.30×10^{-5}	3.587×10^{-8}
Square meter	1550	10.764	1	2.50×10^{-4}	3.861×10^{-7}
Acre	6.270×10^6	43,560	4047	1	1.56×10^{-3}
Square mile	4.014×10^9	2.788×10^7	2.59×10^6	640	1

Volume Equivalents — Table A-3

Unit	Equivalent					
	in.^{3}	gal	ft^3	m^3	acre-ft	cfs-day
Cubic inch	1	0.00433	5.79×10^{-4}	1.64×10^{-5}	1.33×10^{-8}	6.70×10^{-9}
Gallon	231	1	0.134	0.00379	3.07×10^{-6}	1.55×10^{-6}
Cubic foot	1728	7.48	1	0.0283	2.30×10^{-5}	1.16×10^{-5}
Cubic meter	61,000	264	35.3	1	8.11×10^{-4}	4.09×10^{-4}
Acre-foot	7.53×10^7	3.26×10^5	43,560	1233	1	0.504
Cubic foot per second-day	1.49×10^8	6.46×10^5	86,400	2447	1.98	1

APPENDIX A: UNITS AND CONVERSIONS

Velocity Equivalents Table A-4

Unit	Equivalent				
	ft/sec	mi/hr	m/sec	km/hr	kn
Feet per second	1	0.6818	0.3048	1.097	0.5925
Miles per hour	1.467	1	0.4470	1.609	0.8690
Meters per second	3.281	2.237	1	3.600	1.944
Kilometers per hour	0.9113	0.6214	0.2778	1	0.5400
Knots	1.688	1.151	0.5144	1.852	1

Discharge Equivalents Table A-5

Unit	Equivalent					
	gal/day	ft³/day	gal/min	acre-ft/day	cfs	m³/s
U.S. gallons per day	1	0.134	6.94×10^{-4}	3.07×10^{-6}	1.55×10^{-6}	4.38×10^{-8}
Cubic feet per day	7.48	1	5.19×10^{-3}	2.30×10^{-5}	1.16×10^{-5}	3.28×10^{-7}
U.S. gallons per minute	1440	193	1	4.42×10^{-3}	2.23×10^{-3}	6.31×10^{-5}
Acre-feet per day	3.26×10^5	43,560	226	1	0.504	0.0143
Cubic feet per second	6.46×10^5	86,400	449	1.98	1	0.0283
Cubic meters per second	2.28×10^7	3.05×10^6	15,800	70.0	35.3	1

Pressure Equivalents Table A-6

Unit	Equivalent						
	ft H_2O	in. Hg	mm Hg	mbar	kPa	psi	kg/m²
Foot of water(32°F)	1	0.883	22.42	29.89	2.989	0.4335	304.8
Inch of mercury(32°F)	1.133	1	25.40	33.86	3.386	0.4912	345.3
Millimeter of mercury(0°C)	0.0446	0.03937	1	1.333	0.1333	0.01934	13.60
Millbar	0.0335	0.02953	0.7501	1	0.1000	0.01450	10.20
Kilopascal($N/m^2 \times 10^3$)	0.335	0.2953	7.501	10.00	1	0.1450	102.0
Pounds per square inch	2.307	2.036	51.71	68.95	6.895	1	703.1
Kilograms per square meter	0.000328	0.002896	0.07356	0.09807	0.009807	0.001422	1

Energy Equivalents Table A-7

Unit	Equivalent					
	Btu	cal	J	kW-hr	ft-lb	hp-hr
British thermal unit(60°F)	1	252.0	1055	0.0002930	777.9	0.0003929
Calorie(15°C)	0.003969	1	4.186	1.163×10^{-6}	3.087	1.559×10^{-6}
Joule	0.0009482	0.2389	1	2.778×10^{-7}	0.7376	3.725×10^{-7}
Kilowatt-hour	3413	8,600,100	3.600×10^6	1	2.655×10^6	1.341
Foot-pound	0.001286	0.3239	1.356	3.766×10^{-7}	1	5.051×10^{-7}
Horsepower-hour	2545	641,300	2.685×10^6	0.7457	1.980×10^6	1

Power Equivalents Table A-8

Unit	Equivalent				
	W or J/sec	kW	ft-lb/sec	hp	Btu/hr
Watts(or Joules per second)	1	0.001	0.737	0.00134	3.412
Kilowatts	1000	1	737.6	1.314	3412
Foot-pounds per second	1.356	0.001356	1	0.001818	4.63
Horsepower	745.5	0.7455	550	1	2545
British thermal units per hour	0.293	2.93×10^{-4}	0.216	3.93×10^{-4}	1

Dynamic Viscosity Equivalents Table A-9

Unit	Equivalent			
	N-sec/m^2	g/cm-sec(poise)	lb-sec/ft^2	kg/m-hr
Newtons-seconds per square meter	1	10.0	0.0209	3600
Grams per centimeter-second	0.1	1	2.089×10^{-3}	360
Pounds-seconds per square foot	47.88	478.80	1	1.724×10^5
Kilograms per meter-hour	2.778×10^{-4}	2.778×10^{-3}	5.80×10^{-6}	1

1poise=100centipoise(cp)

APPENDIX B: SI UNIT PREFIXES

Prefix	Symbol	Factor
tera	T	10^{12}
giga	G	10^{9}
mega	M	10^{6}
kilo	k	10^{3}
hecto	h	10^{2}
centi	c	10^{-2}
milli	m	10^{-3}
micro	μ	10^{-6}
nano	n	10^{-9}
pico	p	10^{-12}

APPENDIX C: PHYSICAL PROPERTIES OF WATER (SI UNITS)

Temperature	Specific Gravity	Density	Surface Tension	Heat of Vaporization	Viscosity		Bulk Modulus of Elasticity	Vapor Pressure		
					Dynamic	Kinematic				
(°C)		(g/cm^3)	(N/m)	(cal/g)	(poise)	(stokes)	(N/m^2)	Hg	kPa	g/cm^2
0	0.99987	0.99984	75.6×10^{-3}	597.3	1.79×10^{-2}	1.79×10^{-2}	2.02×10^9	4.58	0.611	6.23
5	0.99999	0.99996	74.9	594.5	1.52	1.52	2.06	6.54	0.873	8.89
10	0.99973	0.9997	74.2	591.7	1.31	1.31	2.1	9.2	1.228	12.51
15	0.99913	0.9991	73.5	588.9	1.14	1.14	2.14	12.78	1.706	17.38
20	0.99824	0.99821	72.8	586	1	1	2.18	17.53	2.337	23.83
25	0.99708	0.99705	72	583.2	0.89	0.893	2.22	23.76	3.169	32.3
30	0.99568	0.99565	71.2	580.4	0.798	0.801	2.25	31.83	4.243	43.27
35	0.99407	0.99404	70.4	577.6	0.719	0.723	2.27	42.18	5.625	57.34
40	0.99225	0.99222	69.6	574.7	0.653	0.658	2.28	55.34	7.378	75.23
50	0.98807	0.98804	67.9	569	0.547	0.554	2.29	92.56	12.34	125.83
60	0.98323	0.9832	66.2	563.2	0.466	0.474	2.28	149.46	19.926	203.19
70	0.9778	0.97777	64.4	557.4	0.404	0.413	2.25	233.79	31.169	317.84
80	0.97182	0.97179	62.6	551.4	0.355	0.365	2.2	355.28	47.367	483.01
90	0.96534	0.96531	60.8	545.3	0.315	0.326	2.14	525.89	70.113	714.95

APPENDIX D: GLOSSARY

A

actual evaporation
Evaporation which occurs at a rate controlled by the available water (e.g. plant transpiration may be restricted by low soil moisture).

alkalinity
A measure of the capacity to absorb hydrogen ions without a change in pH. This is influenced by the concentration of hydroxide, bicarbonate or carbonate ions.

aquifer
A layer of unconsolidated or consolidated rock that is able to transmit and store enough water for extraction. A confined aquifer has restricted flow above and below it while an unconfined aquifer has no upper limit.

aquifuge
A totally impermeable rock formation.

aquitard
A geological formation that transmits water at a much slower rate than the aquifer (similar to aquifuge).

areal rainfall
The average rainfall for an area (often a catchment in hydrology) calculated from several different point measurements.

artesian water or well
Water that flows directly to the surface from a confined aquifer (i.e. it does not require extraction from the ground via a pump). The water in aquifer is under pressure so it is able to reach the surface of a well.

AVHRR (Advanced Very High Resolution Radiometer)

A North American Space Agency (NASA) satellite used mainly for atmospheric interpretation.

B
baseflow

The portion of streamflow that is not attributed to storm precipitation (i.e. it flows regardless of the daily variation in rainfall). Sometimes also referred to as slowflow.

Bergeron process

The process of raindrop growth through a strong water vapor gradient between ice crystals and small water droplets.

Boyle's Law

A law of physics relating pressure (P), temperature (T), volume (V) and concentration of molecules (n) in gases.

C
canopy storage capacity

The volume of water that can be held in the canopy before water starts dripping as indirect throughfall.

capillary forces

The forces holding back soil water so that it does not drain completely through a soil under gravity. The primary cause of capillary forces is surface tension between the water and soil surfaces.

catchment

The area of land from which water flows towards a river and then from that river to the sea. Also known as a river basin or a watershed.

channel flow

Water flowing within a channel. A general term for streamflow or riverflow.

channelization

The confinement of a river into a permanent, rigid, channel structure. This often occurs as part of urbanization and flood protection.

cloud seeding

The artificial generation of precipitation (normally rainfall) through provision of extra condensation nuclei within a cloud.

condensation

The movement of water from a gaseous state into a liquid state; the opposite of evaporation.

condensation nuclei

Minute particles present in the atmosphere upon which the water or ice droplets form.

convective precipitation

Precipitation caused by heating from the Earth's surface (leading to uplift of a moist air body).

cyclonic precipitation

Precipitation caused by a low-pressure weather system where the air is constantly being forced upwards.

D
dewfall (or dew)

Water that condenses from the atmosphere (upon cooling) onto a surface (frequently vegetation).

dilution gauging

A technique to measure streamflow based on the dilution of a tracer by the water in the stream.

discharge

In hydrology, discharge is frequently used to denote the amount of water flowing down a river/stream with time (unit: m^3/s, called cumecs).

E
effective rainfall

The rainfall that produces stormflow (i.e. it is not absorbed by soil). This is a term used in the derivation and implementation of the unit hydrograph.

evaporation

The movement of water from a liquid to a gaseous form (i.e. water vapor) and dispersal into the atmosphere.

evaporation pan

A large vessel of water, with a measuring instrument or weighing device underneath that allows you to record how much water is lost through evaporation over a time period.

evapotranspiration

A combination of direct evaporation from soil/water and transpiration from plants. The term recognizes the fact that much of the Earth's surface is a mixture of vegetation cover and bare soil.

F
field capacity

The actual maximum water content that a soil can hold under normal field conditions. This is often less than the saturated water content as the water does not fill all the pore space and gravity drains large pores very quickly.

flash flood

A flood event that occurs as a result of extremely intense rainfall causing a rapid rise in water levels in a stream. This is common in arid and semi-arid regions.

flood

An inundation of land adjacent to a river caused by a period of abnormally large discharge or sea encroachment on the land.

flood frequency analysis

A technique to investigate the magnitude-frequency relationship for floods in a particular river. This is based on historical hydrograph records.

flow duration curve

A graphical description of the percentage of time that a certain discharge is exceeded for a particular river.

flux

The rate of flow of some quantity (e.g. the rate of flow of water as evaporation is referred to as an evaporative flux).

frequency-magnitude

The relationship between how often a particular event (e.g. flood) occurs, and how large the event is. In hydrology it is common to study low frequency-high magnitude events (e.g. large floods do not happen very often).

G

Geographic Information Systems (GIS)

A computer program which is able to store, manipulate and display spatial digital data over an area (e.g. maps).

geomorphology

The study of landforms and how they have evolved.

gravimetric soil moisture content

The ratio of the weight of water in a soil to the overall weight of a soil.

groundwater

Water held in the saturated zone beneath a water table. The area of groundwater is also referred to as water in the phreatic zone.

groundwater flow

Water which moves down a hydraulic gradient in the saturated (phreatic) zone.

H

hillslope hydrology

The study of hydrological processes operating at the hillslope scale.

hydraulic conductivity

A measure of the ability of a porous medium to transmit water. This is a flux term with units of meters per second. The hydraulic conductivity of a soil is highly dependent on water content.

hydrogen bonding

Bonding between atoms or molecules caused by the electrical attraction between a negative and positive ion. This type of bonding exists between water molecules.

hydrograph

A continuous record of streamflow.

hydrograph separation

The splitting of a hydrograph into stormflow and baseflow.

hydrological cycle

A conceptual model of how water moves around between the Earth and atmosphere in different states as a gas, liquid or solid. This can be at the global or catchment scale.

hydrology

"The science or study of" ("logy" from Latin logia) "water" ("hydro" from Greek hudor). Modern hydrology is concerned with the distribution of fresh water on the surface of the Earth and its movement over and beneath the surface, and through the atmosphere.

hydrometry

The science of streamflow measurement.

hydrophobicity and hydrophobic soils

The ability of some soils to rapidly swell upon contact with water so that the initial infiltration rate is low. In this case the water will run over the surface as infiltration excess overland flow.

hypsometric method

A method for estimating areal rainfall based on the topography of the area (e. g. a catchment).

I

infiltration capacity

The rate of infiltration of water into a soil when a soil is fully saturated (i. e. at full capacity of water).

infiltration excess overland flow

Overland flow that occurs when the rainfall rate exceeds the infiltration rate for a soil. Also referred to as Hortonian overland flow.

infiltration rate

How much water enters a soil during a certain time interval.

infiltrometer

An instrument to measure the infiltration rate and infiltration capacity for a soil.

instream flow assessment

A combination of hydrology and aquatic ecology used to assess how much water, and the flow regime, that is required by particular aquatic fauna in a river or stream.

Integrated Catchment Management (ICM)
A form of integrated water resource management (IWRM) that promotes the river catchment as the appropriate organizing unit for understanding and managing water-related biophysical processes in a context that includes social, economic and political considerations. Also sometimes referred to as Integrated Water Basin Management——IWBM.

Integrated Water Resource Management (IWRM)
A water resource management paradigm that promotes the coordinated development and management of water, land and related resources in order to maximize the resultant economic and social welfare in an equitable manner without compromising the sustainability of vital ecosystems.

Interception
The interception of precipitation above the Earth's surface. This may be by a vegetation canopy or buildings. Some of this intercepted water may be evaporated; referred to as interception loss.

isohyetal method
A method for estimating areal rainfall based on the known distribution of rainfall within the area (e. g. a catchment).

K
kriging
A spatial statistics technique that identifies the similarity between adjacent and further afield point measurements. This can be used to interpolate an average surface from a series of point measurements.

L
LANDSAT (LAND Satellite)
A series of satellites launched by the North American Space Agency (NASA) to study the Earth's surface.

low flow
A period of extreme low flow in a river hydrograph (e. g. summer or dry season river flows).

low flow frequency analysis

A technique to investigate the magnitude-frequency relationship for low flows in a particular river. This is based on historical hydrograph records.

M

macropores

Large pores within a soil matrix, typically with a diameter greater than 3mm.

model

A representation of the hydrological processes operating within an area (usually a catchment). This is usually used to mean a numerical model, which simulates the flow in a river, based on mathematical representations of hydrological processes.

mole drainage

An agricultural technique involving the provision of rapid subsurface drainage routes within an agricultural field.

N

net radiation

The total electromagnetic radiation (in all wavelengths) received at a point. This includes direct solar radiation and re-radiation from the Earth's surface.

O

overland flow

Water which runs across the surface of the land before reaching a stream. This is one form (but not the only form) of runoff.

P

peakflow

See "stormflow".

perched water table

Area where the water table is held above a regional water table, usually due to small impermeable lenses in the soil or geological formation.

pH

The concentration of hydrogen ions within a water sample. A measure of water acidity on an inverse logarithmic scale.

phreatic zone

The area beneath a water table (i. e. groundwater).

piezometer

A tube with holes at the base that is placed at depth within a soil or rock mantle to measure the water pressure at a set location.

pipeflow

The rapid movement of water through a hillslope in a series of linked pipes. (NB these can be naturally occurring).

porosity

The percentage of pore space (i. e. air) within a dry soil.

potential evaporation

Evaporation which occurs over the land's surface if the water supply is unrestricted.

precipitation

In hydrology, this is the movement of water from the atmosphere to the Earth's surface. This can occur as rain, hail, sleet or snowfall.

Q
quickflow

See "stormflow".

R
rainfall

Precipitation in a liquid form. The usual expression of rainfall is as a vertical depth of water (e. g. millimeters or inches).

rainfall intensity

The rate at which rainfall occurs. A depth of rainfall per unit time, most commonly in unit of millimeter per hour.

rain gauge

An instrument for measuring the amount of rainfall at a point for a period of time. Standard rain gauges are measured over a day; continuous rainfall measurement can be provided by special rain gauges such as the tipping-bucket gauge.

rain shadow effect

An uneven distribution of rainfall caused by a large high landmass (e.g. a mountain range). On the downwind side of the mountain range there is often less rainfall (i.e. the mountain casts a rain shadow).

rating curve

The relationship between river stage (height) and discharge.

recession limb (of hydrograph)

The period after a peak of stormflow where the streamflow values gradually recede.

rising limb (of hydrograph)

The start of a stormflow peak.

river

A large natural stream of water flowing over the surface and normally contained within a river channel.

river basin

The area of land from which water flows towards a river and then in that river to the sea. Also known as the river catchment.

roughness coefficient

A term used in equations such as Chezy and Manning's to estimate the degree that water is slowed down by friction along the bed surface.

runoff

The movement of liquid water above and below the surface of the Earth prior to reaching a stream or river.

S

salination

The build up of salts in a soil or water body.

satellite remote sensing

The interpretation of ground (or atmospheric) characteristics based on measurements of radiation from the Earth/atmosphere. The radiation measurements are received on satellite-based sensors.

saturated overland flow

Overland flow that occurs when a soil is completely saturated.

saturated water content

The maximum amount of water that the soil can hold. It is equivalent to the soil porosity, which assumes that the water fills all the pore space within a soil.

sensible heat

The heat which can be sensed or felt. This is most easily understood as the heat we feel as warmth. The sensible heat flux is the rate of flow of that sensible heat.

snowfall

Precipitation in a solid form. For hydrology, it is common to express the snowfall as a vertical depth of liquid (i. e. melted) water.

soil moisture deficit

The amount of water required to fill the soil up to field capacity.

soil suction

A measure of the strength of the capillary forces. This is also called the moisture tension or soil water tension. A dry soil exerts a high soil suction.

soil water

Water in the unsaturated zone occurring above a water table. This is also referred to as water in the vadose zone.

specific heat capacity

The amount of energy required to raise the temperature of a substance by a single degree.

SPOT

French satellite to study the Earth's surface.

stage

In hydrology, this term is used to mean the water level height of a river.

stemflow

Rainfall that is intercepted by stems and branches and flows down the tree trunk into the soil.

stomatal or canopy resistance

The restriction a plant places on its transpiration rate through opening and closing stomata in the leaves.

storage

A term in the water balance equation to account for water that is not a flux or is very slowly moving. This may include snow and ice, groundwater and lakes.

storm duration

The length of time between rainfall starting and ending within a storm.

stormflow

The portion of streamflow (normally seen in a hydrograph) that can be attributed to storm precipitation. Sometimes also referred to as quickflow or peakflow.

stream

A small river.

streamflow

Water flowing within a stream channel (or river flow for a larger body of water). Often referred to as discharge.

Synthetic Aperture Radar (SAR)

A remote sensing technique that uses radar properties, usually of microwaves.

synthetic unit hydrograph

A unit hydrograph derived from knowledge of catchment characteristics rather than historical hydrograph records.

T

tensiometer

An instrument used to measure the soil moisture tension.

Thiessen's Polygons

A method of estimating average rainfall for an area based on the spatial distribution of rain gauges.

throughfall

The precipitation that falls to the ground either directly (through gaps in the canopy), or indirectly (having dripped off leaves, stems or branches).

throughflow

Water which runs to a stream through the soils. This is frequently within the unsaturated (vadose) zone. This is one form of runoff. Sometimes referred to as lateral flow.

Time Domain Reflectometry (TDR)

A method to estimate the soil water content based on the interference of propagated electromagnetic waves due to water content.

Total Dissolved Solids (TDS)

The amount of solids dissolved within a water sample. This is closely related to the electrical conductivity of a water sample.

Total Suspended Solids (TSS)

The amount of solids suspended within a water sample. This is closely related to the turbidity of a water sample.

transpiration

The movement of liquid water from a plant leaf to water vapor in the atmosphere. Plants carry out transpiration as part of the photosynthetic process.

turbidity

The cloudiness of a water sample.

U

ultrasonic flow gauge

An instrument that measures stream discharge based on the alteration to a propagated wave over a known cross section.

unit hydrograph

A model of stormflow in a particular catchment used to predict possible future storm impacts. It is derived from historical hydrograph records.

V

vadose zone

Area between the water table and the Earth surface. The soil/rock is normally partially saturated.

vapor pressure
Pressure exerted within the parcel of air by having the water vapor present within it. The more water vapor is present, the greater the vapor pressure will be.

vapor pressure deficit
The difference between actual vapor pressure and saturation vapor pressure.

variable source areas concept
The idea that only certain parts of a catchment area contribute overland flow to stormflow and that these vary in space and time; compare to the "partial areas concept".

velocity-area method
A technique to measure instantaneous streamflow through measuring the cross-sectional area and the velocity through the cross section.

volumetric soil moisture content
The ratio of the volume of water in a soil to the overall volume of a soil.

W

water balance equation
A mathematical description of the hydrological processes operating within a given timeframe. Normally it includes precipitation, runoff, evaporation and change in storage.

water table
The surface that differentiates between fully saturated and partially saturated soil/rock.

water vapor
Water in a gaseous form.

well
A tube with permeable sides all the way up so that water can enter or exit from anywhere up the column. Wells are commonly used for water extraction and monitoring the water table in unconfined aquifers.

wilting point
The soil water content when plants start to die back (wilt).

REFERENCES

[1] BROOKS K N, FFOLLIOTT P F, MAGNER J A. Hydrology and the Management of Watersheds (4th ed) [M]. New York: John Wiley & Sons, 2013.
[2] BRUTSAERT W. Hydrology: An Introduction [M]. London: Cambridge University Press, 2005.
[3] ACREMAN M. The Hydrology of the UK, a Study of Change [M]. London: Routledge, 2000.
[4] DAVIE T. Fundamentals of Hydrology (2nd ed) [M]. Taylor & Francis e-Library, 2008.
[5] CHARBENEAU R J. Groundwater Hydraulics and Pollutant Transport [M]. Upper Saddle River: Prentice Hall, 2000.
[6] FITTS C R. Groundwater Science [M]. Bath: The Bath Press, 2002.
[7] GUPTA R S. Hydrology and Hydraulic Systems (4th ed) [M]. Long Grove: Waveland Press, 2017.
[8] MILLER D. Water at the Surface of the Earth——An Introduction to Ecosystem Hydrodynamics [M]. London: Academic Press, 1977.
[9] DAVIS M L, MASTEN S J. Principles of Environmental Engineering and Science (3rd ed) [M]. New York: McGraw-Hill, 2014.
[10] Intergovernmental Panel on Climate Change (IPCC). Climate change 2007 [M]. Cambridge University Press, 2007, http://www.ipcc.ch.
[11] Global Water Partnership. Catalyzing Change: a Handbook for Developing Integrated Water Resources Management (IWRM) and Water Efficiency Strategies [M]. 2004.
[12] BUDYKO M I. Climate and Life [M]. London: Academic Press, 1974.
[13] KORZUN V I, SOKOLOV A A, BUDYKO M I, KALININ G P. World Water Balance and Water Resources of the Earth, USSR National Committee for the International Hydrological Decade [M]. UNESCO Press, 1978.
[14] HORNBERGER G M, WIBERG P L, RAFFENSPERGER J P, D'ODORICO P. Elements of Physical Hydrology (2nd ed) [M]. Baltimore: Johns Hopkins University Press, 2014.
[15] BRAS R L. Hydrology: An Introduction to Hydrologic Science [M]. Addison-Wesley Publishing, 1990.
[16] National Research Council (NRC). Challenges and Opportunities in the Hydrologic Sciences [M]. Washington, DC: The National Academies Press, 2012.
[17] National Research Council (NRC). Climate Change: Evidence, Impacts, and Choices [M]. The Washington, DC: The National Academies Press, 2012.
[18] HORTON R E. Erosional development of streams and their drainage basins; hydrophysical approach to quantitative morphology [J]. Geological Society of America Bulletin, 1945, 56 (3): 807-813.
[19] SCHUMM S A. The Fluvial System [M]. New York: John Wiley & Sons, 1977.
[20] STRAHLER A N. Quantitative Geomorphology of Drainage Basins and Channel Networks. In Handbook of Applied Hydrology [M]. New York: McGraw-Hill, 1964.
[21] BONAN G. Ecological Climatology [M]. Cambridge: Cambridge University Press, 2008.
[22] BOYER E W, HORNBERGER G M, BENCALA K E, MCKNIGHT D M. Response characteristics of DOC flushing in an alpine catchment [J]. Hydrological Processes, 1997, 11 (12): 1635-1647.
[23] LINSLEY R K, FRANZINI J B, FREYBERG D L, TCHOBANOGLOUS G. Water Resources Engineering (4th ed) [M]. New York: McGraw-Hill, 1992.
[24] FETTER C W. Applied Hydrogeology (4th ed) [M]. New York: Macmillan, 2000.
[25] RAO S N. Hydrogeology: Problems with Solutions [M]. Delhi: PHI Learning Private Limited, 2016.
[26] BOUWER H, Groundwater Hydrology [M]. New York: McGraw-Hill, 1978.
[27] CHOW V T. Handbook of Applied Hydrology [M]. New York: McGraw-Hill, 1964.
[28] DAVIS S N, DEWIEST R J M. Hydrogeology [M]. New York: John Wiley & Sons, 1967.
[29] RAGHUNATH H M. Hydrology Principles, Analysis and Design [M]. New Delhi: New Age

International, 2015.
[30] RAYMAHASHAY B C. Geochemistry for Hydrologists [M]. New Delhi: Allied Publishers, 1996.
[31] SUBRAHMANYAM V P. Water Balance and its Applications [M]. Visakhapatnam: Andhra University Press, 1982.
[32] TODD D K. Groundwater Hydrology [M]. New York: John Wiley & Sons, 2006.
[33] THOMAS H C. Low Impact Development and Sustainable Stormwater Management [M]. Hoboken: John Wiley & Sons, 2012.
[34] FREEZE R A, CHERRY J A. Groundwater [M]. Englewood Cliffs: Prentice Hall, 1979.
[35] NEUZIL C E. How permeable are clays and shales? [J]. Water Resources Research, 1994, 30 (2), 145-150.
[36] BEDIENT P, HUBER W C, VIEUX B E. Hydrology and Floodplain Analysis (4th ed) [M]. Pearson, 2008.
[37] MILLER G T, SPOOLMAN S E. Environmental Science (14th ed) [M]. Belmont: Yolanda Cossio, 2013.
[38] ANDERSON M G, MCDONNELL J J. Encyclopedia of Hydrological Sciences [M]. Chichester: John Wiley & Sons, 2006.
[39] LI S, KAZEMI H, ROCKAWAY T D. Performance assessment of stormwater GI practices using artificial neural networks [J]. Science of the Total Environment, 2019, 651: 2811-2819.
[40] FREEZE A R, WITHERSPOON A P. Effect of water-table configuration and subsurface permeability variation [J]. Water Resources Research, 1967, 3 (2): 623-634.
[41] TURNIPSEED D P, SAUER V B. Discharge Measurements at Gaging Stations [R]. Reston: U. S. Geological Survey, 2010.
[42] VAN DER LEEDEN F, TROISE F L, TODD DK. The Water Encyclopedia (2nd ed) [M]. Boca Raton: Lewis Publishers, 1990.